Cyber-physical Systems

Cyber-physical Systems

Theory, Methodology, and Applications

Pedro H. J. Nardelli

IEEE PRESS

WILEY

For general information on our other products and services or for technical support, please contact our Customer Care Department within the United States at (800) 762-2974, outside the United States at (317) 572-3993 or fax (317) 572-4002.

Wiley also publishes its books in a variety of electronic formats. Some content that appears in print may not be available in electronic formats. For more information about Wiley products, visit our web site at www.wiley.com.

Library of Congress Cataloging-in-Publication Data applied for:
ISBN: 9781119785163

Cover Design and Image: Wiley

Set in 9.5/12.5pt STIXTwoText by Straive, Chennai, India

10 9 8 7 6 5 4 3 2 1

Contents

Preface

This book has been written as an introductory text to cyber-physical systems conceptualized as a scientific object in itself. The decision to move in this direction was based on my subjective experience of the objective reality of engineering education and academic research, very often overspecialized and too modular. I tried to go beyond such a reductionist view by making evident the complex articulations that constitute cyber-physical systems as such, as well as the necessary relations to the external world in which they are – or will be – deployed, being them physical or social. My advise to anyone who wants to learn or teach from this book is (i) to follow the chapters in the order they are presented, (ii) do the proposed exercises, and (iii) keep the mind open to understand the articulation of concepts that define the presented theory. Only in this way, the reader or educator can fully enjoy the strength of the theory as the basis of a methodological framework for practical interventions.

The history of the manuscript is the following. The very first, preliminary version of the manuscript-to-be was presented as tutorial notes in the *2017 International Symposium on Wireless Communication Systems* in Bologna (Italy). Then, this tutorial text was extended to become the lecture notes of a completely new course I had the freedom to develop at LUT as soon as I moved from the University of Oulu to LUT in 2018; the course is called *Introduction of IoT-based Systems*. In 2020, I decided to convert those notes into a real book, which Wiley kindly accepted to publish.

I would like to acknowledge all the colleagues with whom I have discussed topics related to this book that in some way or another have helped to shape it, specially the friends at LUT University (Finland), the University of Oulu (Finland), and the University of Campinas (Brazil). Among those persons some deserve special praise, namely (i) Dr. Hanna Niemelä – an associate professor at LUT – for proofreading the whole manuscript and also giving suggestions, (ii) Arthur Sena, M.Sc. – a doctoral student at LUT – for drawing so many figures and helping me in some technical parts of the book, (iii) Dr. Alysson Mascaro – professor of philosophy of law at the University of São Paulo, Brazil – for guiding my studies in critical Marxist philosophy, in particular Althusser, and (iv) Dr. Harun Siljak – assistant professor of embedded systems, optimization and control in Trinity College Dublin, Ireland – for sharing so many ideas about cyber-physical systems, cybernetics, and information theory (mostly in a revolutionary way). I would also like to thank Dr. Florian Kühnlenz, who at some point in 2014 talked with me about one of my schematic handmade drafts that were randomly placed on my messy table and a few days later presented to me quite an interesting simulation where he implemented those ideas. This was the beginning of his doctoral research at the University of Oulu, which ended up motivating me to dig into the conceptualization of cyber-physical systems that is now presented here.

The research work systematized in this book has been partly supported by (i) the Academy of Finland through the research fellowship project *Building the Energy Internet as a large-scale IoT-based cyber-physical system that manages the energy inventory of distribution grids as discretized packets via machine-type communications* (EnergyNet; grant no. 321265/no. 328869), through the consortium *Framework for the identification of rare events via machine learning and IoT networks* (FIREMAN; grant no. 326270 under CHIST-ERA-17-BDSI-003), and the project *Energy efficient IoT deployment for systems of the future* (ee-IoT; grant no. 319009), (ii) the Jane and Aatos Erkko Foundation (Finland) through the project *Swarming technology for reliable and energy-aware aerial missions* (STREAM), and (iii) through the LUT research platform *Modeling reality through simulation* (MORE SIM).

I dedicate this book to my mom Regina, my dad Eliseu, my grandma Celina, my wife Carolina, and my daughter Amanda.

November 12, 2021   *Pedro H. J. Nardelli*
                                                                                              Lappeenranta

1

Introduction

What is a cyber-physical system? Why should I study it? What are its relations to cybernetics, information theory, embedded systems, industrial automation, computer sciences, and even physics? Will cyber-physical systems (CPSs) be the seed of revolutions in industrial production and/or social relations? Is this book about theory or practice? Is it about mathematics, applied sciences, technology, or even philosophy? These are the questions the reader is probably thinking about right now. Definitive answers to them are indeed hard to give at this point. During the reading of this book, though, I expect that these questions will be systematically answered. Hence, new problems and solutions can then be formulated, allowing for a progressive development of a new scientific field.

As a prelude, this first chapter will explicitly state the philosophical position followed in this book. The chapter starts by highlighting what is spontaneously thought under the term "CPS" to then argue the reasons why a general theory is necessary to build scientific knowledge about this object of inquiry and design. A brief historical perspective of closely related fields, namely control theory, information theory, and cybernetics, will also be given followed by a necessary digression of philosophical positions and possible misinterpretations of such broad theoretical constructions. In summary, the proposed demarcation can be seen as a risk management action to avoid mistakes arising from commonsense knowledge and other possible misconceptions in order to "clear the path" for the learning process to be carried out in the following hundreds of pages.

1.1 Cyber-Physical Systems in 2020

Two thousand and twenty is a remarkable year, not for the high hopes the number 20-20 brought, but for the series of critical events that have happened and affected everyone's life. The already fast-pace trend of digitalization, which had started decades before, has boomed as a consequence of severe mobility restrictions

Cyber-physical Systems: Theory, Methodology, and Applications, First Edition. Pedro H. J. Nardelli.
© 2022 The Institute of Electrical and Electronics Engineers, Inc. Published 2022 by John Wiley & Sons, Inc.

imposed as a response to the COVID-19 pandemics. The uses (and abuses) of information and communication technologies (ICTs) are firmly established and widespread in society. From dating to food delivery, from reading news to buying e-books, from watching youtubers to arguing through tweets, the *cyber* world – before deemed in science fiction literature and movies as either utopian or apocalyptic – is now very concrete and pervasive. Is this concreteness of all those practices involving computers or computer networks (i.e. cyber-practices) what defines *CPSs*? In some sense, yes; in many others, no; it all depends on how CPS is conceptualized! At all events, let us move step-by-step by looking at nonscientific definitions.

CPS is a term not broadly employed in everyday life. Its usage has a technical origin and is related to digitalization of processes across different sectors so that the term "CPS" has ended up being mostly used by academics in information technology, engineering, practitioners in industry, and managers. Such a broad concept usually leads to misunderstandings so much so that relevant standardization bodies have channeled efforts trying to establish a shared meaning. One remarkable example is the *National Institute of Standards and Technology* (NIST) located in the United States. NIST has several working groups related to CPS, whose outcomes are presented on a dedicated website [1]. In NIST's own words,

> Cyber-Physical Systems (CPS) comprise interacting digital, analog, physical, and human components engineered for function through integrated physics and logic. These systems will provide the foundation of our critical infrastructure, form the basis of emerging and future smart services, and improve our quality of life in many areas.
>
> Cyber-physical systems (CPS) will bring advances in personalized health care, emergency response, traffic flow management, and electric power generation and delivery, as well as in many other areas now just being envisioned. CPS comprise interacting digital, analog, physical, and human components engineered for function through integrated physics and logic. Other phrases that you might hear when discussing these and related CPS technologies include:
>
> - Internet of Things (IoT)
> - Industrial Internet
> - Smart Cities
> - Smart Grid
> - "Smart" Anything (e.g. Cars, Buildings, Homes, Manufacturing, Hospitals, Appliances)

As a commonplace when trying to determine the meaning of umbrella terms, the definition of CPS proposed by NIST is still too broad and vague (and excessively utopian) to become susceptible of scientific inquiry. On the other hand, such a

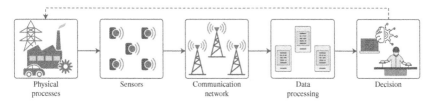

| Physical processes | Sensors | Communication network | Data processing | Decision |

Figure 1.1 Illustration of a CPS. Sensors measure physical processes, whose data are transmitted through a communication network. These data are then processed to support decisions related to the physical process by either a human operator or an expert system.

definition offers us a starting point, which can be seen as the raw material of our theoretical investigation. A careful reading of the NIST text indicates the key common features of the diverse list of CPSs:

- There are physical processes that can be digitalized with sensors or measuring devices;
- These data can be processed and communicated to provide information of such processes;
- These informative data are the basis for decisions (either by humans or by machines) of possible actions that are capable of creating "smartness" in the CPS;
- CPSs are designed to intervene (improve) different concrete processes of our daily lives; therefore, they affect and are affected by different aspects of society.

These points indicate generalities of CPSs, as illustrated in Figure 1.1. In most of the cases, though, they are only *implicitly* considered when particular solutions are analyzed and/or designed. As a matter of fact, specific CPSs do exist in the real world without the systematization to be proposed in this book. So, there is an apparent paradox here: on the one hand, we would like to build a scientific theory for CPSs in general; on the other hand, we see real deployments of particular CPSs that do not use such a theory. The next section will be devoted to resolve this contradiction by explaining the reasons why a general theory for CPS is necessary while practical solutions do indeed exist.

1.2 Need for a General Theory

The idea of having a *general theory* is, roughly speaking, to characterize in a nonsubjective manner a very well-defined symbolic object that incorporates all the constitutive aspects of a class of real-world objects and therefrom obtain new knowledge by both symbolic manipulation and experimental tests. This generalization opens the path for moving beyond know-how-style of knowledge toward abstract, scientific conceptualizations, which are essential to assess existing objects, design new ones, and define their fundamental limits.

Example 1.1 *Chocolate cake.* You would like to prepare the chocolate cake you ate in your childhood, whose recipe you found in the Internet. This instructs you *how* to make it. If you follow the instructions, you will have the cake you wish but without knowing *why* that specific combination of elements and the processes of mixing and cooking can produce such a cake. If you decide to investigate the recipe, you will find that different elements are compatible for chemical reasons, and, after processing then in a specific way, such a combination will have the desired flavor and structural characteristic. Based on this abstraction, you can now (i) understand the reasons why the recipe works, (ii) propose modifications to the recipe to make the cake fluffier or moister, and (iii) create a vegan version of the cake by finding replacement for milk, butter, and eggs.

This example serves as a very simplified illustration of the difference between technical knowledge (know-how) and scientific knowledge. We are going to discuss sciences and scientific practice in more detail when pinpointing the philosophical position taken throughout this book. At this point, though, we should return to our main concern: the need for a general theory that conceptualizes CPSs. Like the particular chocolate cake, the existence of *smart* grids or cities, or even fully automated production lines neither precludes nor requires a general theory. Actually, their existence, the challenges in their particular deployments, and their specific operation can be seen as the necessary raw material for the scientific theory that would build the knowledge of CPSs as a symbolic (general, abstract) object. This theory would provide the theoretical tools for orienting researchers, academics, and practitioners with objective knowledge to analyze, design, and intervene in particular (practical) realizations of this symbolic object called CPS.

Without advancing too much too soon, let us run a thought experiment to mimic a specific function of *smart* meters as part of the *smart* electricity grid – one of the most well-known examples of CPS. Consider the following situation: the price of electricity in a household is defined every hour and the smart meter has access to this information. The smart meter also works as a home energy management system, turning on and off some specific loads or appliances that have flexibility in their usage as, for instance, the washing machine or the charging of an electric vehicle (EV). If there is no smartness in the system, whenever the machine is turned on or the EV is plugged in, they will draw electric energy from the grid. With the smart meter deciding when the flexible load will turn on or off based on the price, the system is expected to become smart overall: not only flexible loads could be turned on when there is a low price (leading to lower costs to the households) but this would also help the grid operation by flattening the electricity demand curve (which has peaks and valleys of consumption ideally reflected by the price).

This seems too good to be true, and it indeed is! The trick is the following. The electricity price is a *universal signal* so that all smart meters see the same value. By facing the same price, the smart meters will tend to switch on (and then off) the loads at the same time, creating a collective behavior that would probably lead to unexpected new peaks and valleys in demand that cannot be predicted by the price (which actually reflects past behaviors, and not instantaneous ones). The adage *the whole is more than the sum of its parts* then acquires a new form: the smartness of the smart meters can potentially yield a stupid grid [2]. This outcome is surprising since individually everything is working as expected, but the system-level dynamic is totally undesirable. How to explain this?

The smart grid, as we have seen before, is considered a CPS where physical processes related to energy supply and demand are reflected in the cyber domain by a price signal that serves as the basis for the decisions of smart meters, which then modify the physical process of electricity demand by turning on appliances. However, the smart meters described above are designed to operate considering the grid dynamics as given so that they individually react to the price signal assuming that they cannot affect the electricity demand at the system level. If several of such smart meters operate in the grid by reacting to the same price signals, they will tend to have the same decisions and, consequently, coordinate their actions, leading to the undesirable and unexpected aggregate behavior. This is a byproduct of a segmented way of conceptualizing CPSs, which overestimates the smartness of devices working individually while underestimating the physical and logical (cyber) interrelations that constitute the smartness of the CPS.

Figure 1.2 illustrates this fact. The electricity hourly price is determined from the expected electricity demand following the indication of arrow (a). The smart meters would lead to a smart grid if the spikes in consumption would be flattened, as indicated by arrow (b). However, due to the unexpected collective effect

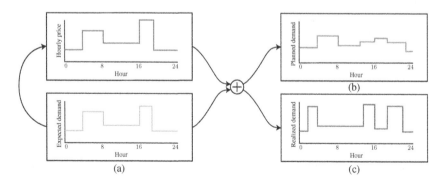

Figure 1.2 Operation of smart meters that react to hourly electricity price. Arrow (a) indicates that the hourly price is a function of the expected electricity demand. Arrow (b) represents the planned outcome of demand response, namely decreasing spikes. Arrow (c) shows the realized demand as an aggregate response to price signals, leading to more spikes.

of reactions to the hourly price, the realized demand has more spikes than before, creating a stupid grid. This is pointed by arrow (c).

At all events, coordination and organization of elements working together are old problems under established disciplines like systems engineering and operational research. However, those disciplines are fundamentally based on centralized decision-making and optimal operating points; this is usually called a top-down approach. In CPSs, there is an internal awareness and a possibility of distributed or decentralized decision-making based on local data and (pre-defined or learned) rules co-existing with hierarchical processes. Hence, CPSs cannot be properly characterized without explicit definition of how data are processed, distributed, and utilized for informed decisions and then actions. A general (scientific) theory needs to be built upon the facts that CPSs have internal communication–computation structures with specific topologies that result in internal actions based on potentially heterogeneous decision-making processes that internally modify the system dynamics. Some of these topics have been historically discussed within control and information theories, as well as cybernetics.

1.3 Historical Highlights: Control Theory, Information Theory, and Cybernetics

Control theory refers to the body of knowledge involving automatic mechanisms capable of self-regulating the behavior of systems. Artifacts dating back to thousands of years indicate the key idea behind feedback control loop. The oldest example is probably the *water clock* where inflow and outflow of water is used to measure time utilizing a *flow regulator*, whose function is [3]: "(…) to keep the water level in a tank at a constant depth. This constant depth yielded a constant flow of water through a tube at the bottom of the tank which filled a second tank at a constant rate. The level of water in the second tank thus depended on time elapsed." Other devices based on the same principle have also been found throughout history. With the industrial revolution in the 1700s, different types of *regulators* and *governors* have appeared as part of the unprecedented technological development. It was desirable for the new set of machinery like windmills, furnaces, and steam engines to be controlled automatically. Hence, more and more solutions based on feedback and self-regulation started to be developed. At that point, the development was of a practical nature, by trial-and-error. In a groundbreaking work by J. C. Maxwell (also the "father" of electromagnetism) from 1868 [4], the first step to generalize feedback control as a scientific theory was taken. Maxwell analyzed systems that employ governors by linearized differential equations to establish the necessary conditions for stability based on the system's characteristic equation. This level of generalization that states fundamental (mathematical) conditions that apply to all existing and potentially future objects of the same kind (i.e. feedback control systems) is the heart of the scientific endeavor of *control theory*.

A similar path can be also seen with information theory. Data acquisition, processing, transmission, and reception are biological facts, not only in humans but in other animals. Without the trouble to define what information is now, it is clear that different human societies have come up with different ways of sending informative data from one point to another. From spoken language to books, from smoke signals to pigeons, data can be transferred in space and time, *always facing the possibility of error*. With the radical changes brought by the industrial revolution, quicker data transmission across longer distances was desired [5]. The first telegraphs appeared in the early 1800s; the first transatlantic telegraph dates to 1866. By the end of the 19th century, the first communication networks were deployed in the United States. Interestingly enough, the development of communications networks depended on amplifiers and the use of feedback control, which have been studied in control theory.

Even with a great technological development of communication networks, mainly carried out within the Bell Labs, transmission errors had always been considered inevitable. Most solutions were focusing on how to decrease the chances of such events, also in a sort of highly complex trial-and-error fashion. In 1948, in one edition of *The Bell System Technical Journal*, C. E. Shannon published one amazing piece of work stating in a fully mathematical manner the fundamental limits of communication systems utilizing a newly proposed definition of information based on *entropy*; some more details will come later in Chapter 4. What is important here is to say that, in contrast to all previous contributions, Shannon created a mathematical model that states the fundamental limits of *all* existing and possible communication systems by determining the capacity of communication channels. One of his key results was the counter-intuitive statement that *an error-free transmission is possible if, and only if, the communication rate is below the channel capacity*. Once again, one can see Shannon's paper as the birth of a scientific theory of this new well-determined object referred to as *information*.

In the same year that Shannon published *A Mathematical Theory of Communication*, another well-recognized researcher – Norbert Wiener – published a book entitled *Cybernetics: Or Control and Communication in the Animal and the Machine* [6]. This book introduces the term *cybernetics* in reference to self-regulating mechanisms. In his erudite writing, Wiener philosophically discussed several recent developments of control theory, as well as preliminary thoughts on information theory. He presented astonishing scientific-grounded arguments to draw parallels between human-constructed self-regulating machines, on the one side, and animals, humans, social, and biological processes, on the other side. Here, I would like to quote the book *From Newspeak to Cyberspeak* [7]:

> Cybernetics is an unusual historical phenomenon. It is not a traditional scientific discipline, a specific engineering technique, or a philosophical doctrine, although it combines many elements of science, engineering, and philosophy. As presented in Norbert Wiener's classic 1948 book Cybernetics, or Control and Communication in the Animal and the Machine, cybernetics comprises an assortment of analogies between humans and self-regulating machines: human behavior is compared to the operation of a servomechanism; human communication is likened to the transmission of signals over telephone lines; the human brain is compared to computer hardware and the human mind to software; order is identified with life, certainty, and information; disorder is linked to death, uncertainty, and entropy. Cyberneticians view control as a form of communication, and communication as a form of control: both are characterized by purposeful action-based on information exchange via feedback loops. Cybernetics unifies diverse mathematical models, explanatory frameworks, and appealing metaphors from various disciplines by means of a common language that I call cyberspeak. This language combines concepts from physiology (homeostasis and reflex), psychology (behavior and goal), control engineering (control and feedback), thermodynamics (entropy and order), and communication engineering (information, signal, and noise) and generalizes each of them to be equally applicable to living organisms, to self-regulating machines, and to human society.
>
> In the West, cybernetic ideas have elicited a wide range of responses. Some view cybernetics as an embodiment of military patterns of command and control; others see it as an expression of liberal yearning for freedom of communication and grassroots participatory democracy. Some trace the origins of cybernetic ideas to wartime military projects in fire control and cryptology; others point to prewar traditions in control and communication engineering. Some portray cyberneticians' universalistic aspirations as a grant-generating ploy; others hail the cultural shift resulting from cybernetics' erasure of boundaries between organism and machine, between animate and inanimate, between mind and body, and between nature and culture.

We can clearly see a difference between the generality of information and control theories with respect to their own well-defined objects, and the claimed universality of cybernetics that would cover virtually all aspects of reality. In this sense, the first two can be claimed to be scientific theories in the strong sense. The last, despite its elegance, seems less a science but more a theoretical (philosophical) displacement or distortion of established scientific theories by expanding their reach towards other objects. This is actually a very controversial argument that depends on the philosophical position taken throughout this book, whose details will be presented next.

1.4 Philosophical Background

Science is a special type of formal discourse that claims to hold objective true knowledge of well-determined objects. Different sciences have different objects, requiring different methods to state the truth value of different statements. A given science is presented as a *theory* (i.e. a systematic, consistent discourse) that articulates different *concepts* through a chain of *determinations* (e.g. causal or structural relations) that are *independent of any agent (subject)* involved in the production of scientific knowledge. This, however, does not preclude the importance of scientists: they are the necessary *agents* of the scientific practice. Scientific practice can then be thought as the way to produce new knowledge about a given object, where scientists work on theoretical raw material (e.g. commonsense knowledge, know-how knowledge, empirical facts, established scientific knowledge) following historically established norms and methods in a specific scientific field to produce new scientific knowledge. In other words, *scientific practice is the historically defined production process of objective true knowledge.* Note that these norms, despite not being fixed, have a relatively stable structure since the object itself constrains which are the valid methods eligible to produce the knowledge effect.

Moreover, scientific knowledge poses general statements about its object. Such a generality comes with abstraction, moving from particular (narrow) abstractions of real-world, concrete objects to abstract, symbolic ones. Particular variations of a class of concrete objects can be used as the raw material by scientists to build a general theory that is capable of covering all, known and unknown, concrete variations of that class of objects. This general theory is built upon abstract objects that provide knowledge of concrete objects. However, this differentiation is of key importance since a one-to-one map between the concrete and abstract realities may not exist. Abstract (symbolic) objects as part of scientific theories produce a knowledge effect on concrete objects, understood as realizations of the theory, not as a reduction or special case. At any rate, despite the apparent preponderance of abstractions, the concrete reality is what determines in the last instance the validity of the theory (even in the "concrete" symbolic reality of pure mathematics, concreteness is defined by the foundational axioms and valid operations).

To illustrate this position, let us think about dogs. Although the concept of dog cannot bark, dogs do bark. Clearly, in the symbolic reality in which the concept of dog exists, it has the ability of barking. The concept, though, cannot transcend this domain so we cannot hear in the real world the barking sound of the abstracted dog. Conversely, we all hear real dogs barking, and therefore, any abstraction of dogs that assumes that they cannot bark shall not be considered scientific at all. This seems trivial when presented with this naive example, but we will see throughout this book the implications of unsound abstractions in different, more elusive domains. This is even more critical when incorrect abstractions are accompanied by heavily mathematized (therefore consistent) models. For instance, the fact that some statement is a true knowledge in mathematical sciences does not imply it is true in economics. Always remember: a mathematically consistent model is not synonymous with a scientific theory.

Philosophy, like science, is also a theoretical discourse but with a very important difference: it works by demarcating positions as correct or incorrect based on its own philosophical system that defines *categories* and their relations [8]. Unlike scientific proofs, philosophy works through rational argumentation to defend positions (i.e. *theses*), usually trying to answer universal and timeless questions about, for example, existence of freedom. In this case, philosophy has no specific (concrete) object as sciences do; consequently, it is not a science in the way we just defined. Philosophy then becomes its own practice: rational argumentation based on a totalizing system of categories defining positions about everything that exists or not. Following this line of thought, philosophy is not a science of sciences; it can neither judge the truth value of propositions internally established by the different sciences nor state *de jure* conditions for scientific knowledge from the outside.

Besides, scientific and philosophical practices exist among several other social practices. They are part of an articulated historical social whole, where different practices coexist and interfere with each other at certain degrees and levels of effectivity. As previously discussed, scientific practice produces general objective true knowledge about abstract objects, which very usually contradicts the commonsense ideas that are usually related to immediate representations arising from other social practices of our daily lives. This clearly leads to obstacles to scientists, who are both agents of the scientific practice and individuals living in society. The totalizing tendency of philosophy also plays a role: it either *distorts* scientific theories and concepts to fit in universal systems of philosophical categories or *judges* their truth value based on universal methodological assumptions. This directly or indirectly affects the self-understanding of the relation that the scientists have with their own practice, creating new obstacles to the science development [9, 10].

Example 1.2 *Differences between daily language, philosophical categories, and scientific concepts.* One word that exemplifies very well the difference is *time*. We use word *time* in many ways in our daily lives: to discuss about our activities, plans, routine, and the like. However, in philosophy, the category *Time* has different roles depending on the philosophical system to be considered – this usually comes with the relation between other categories like Causality, Origin, and End. In sciences, *time* is also a concept in different disciplines. In physics, *time* is a very precise concept that has been changing throughout its history, changing (not without pain) from the classical definition that time is an absolute measure (i.e. the same everywhere) to today's relativity theory where time is relative (and the speed of light is absolute). Such a scientific definition of the concept of time is not intuitive at all, and goes against most of our immediate use of the word. In this sense, scientists may find it difficult to operate with the scientific concept of time in relativity theory because of the other more usual meanings of the word. Besides, such a confusion between the scientific and the nonscientific may open philosophical questions and nonscientific interpretations of the scientific results.

In addition to this unavoidable challenge, the rationalization required by scientific theories appears in different forms. In this case, philosophical practice can help scientific practice by classifying the different types of rationality depending on the object under consideration. Motivated by Lepskiy [11] (but understood here in a different manner) and Althusser [8], we propose the following division.

- **Classical scientific rationality:** Direct observations and empirical falsification are possible for all elements of the theory, i.e. there is a one-to-one map between the physical and abstract realities.
- **Nonclassical scientific rationality:** Observations are not directly possible, i.e. the process of abstraction leads to nonobservable steps, resulting in a relatively autonomous theoretical domain.
- **Interventionist scientific rationality:** Active elements with internal awareness with objectives and goals exist, leading to a theory of the *fact to be accomplished* in contrast to theories of the *accomplished facts*.

By acknowledging the differences between these forms of rationality, sciences and scientific knowledge can be internalized as a social practice within the existing mode of production. Different from positivist and existentialist traditions in philosophy, this practice of philosophy attempts to articulate the scientific practice within the historical social whole, critically building demarcations of the correctness of the reach of scientific knowledge by rational argumentation [10].

Example 1.3 *Scientific efforts related to COVID-19.* In 2020, an unprecedented channeling of research activities was directed to combat the COVID-19 pandemics. These activities can be classified following the three aforementioned scientific rationalities. The **classical rationality** can be exemplified by the process to test the effectiveness of vaccines following the historically defined norms. The **nonclassical rationality** incorporates the mathematical models for epidemics based on nonlinear dynamical processes over graphs where not all variables are observable. The **interventionist rationality** considers lockdown policies to control the virus propagation as a fact to be accomplished. A critical philosophical practice demarcates the reach of the three different scientific activities, determining both their interrelations and the articulation with other social practices. For example, a vaccine that is proved to work can be modeled by a mathematical model, which can be used to change the lockdown policy. However, within the capitalist mode of production, these activities are directly or indirectly determined by the economical reality – from the funds available to develop the vaccine and its respective property rights to the economic impact of lockdown policies and its justifications based on a wide range of epidemiological models. A critical philosophical practice acknowledges the autonomy of the results obtained through the scientific practice with respect to its object while it internalizes such a practice in the articulated social whole.

This philosophical practice goes hand in hand with the scientific practice by helping scientists to avoid overreaching tendencies related to their own theoretical findings. It also indicates critical points where other practices might be interfering in the scientific activity and vice versa. Although a deep discussion of the complex relations between scientific practice and other practices are far beyond our aim here, we will throughout this book deal with one specific relation: how scientific practice is related to the technological development. We have seen so far that the practical development of techniques does not require the intervention of (abstract) scientific rationalization. On the other hand, the knowledge produced by the sciences has a lot to offer to practical techniques. The existence of the term *technology*, referring to techniques developed or rectified by the sciences, indicates such a relation. More than what this definition might suggest, technology cannot be simply reduced to a mere application of scientific knowledge; it can indeed create new domains and objects subject to a new scientific discourse.

The aforementioned control and information theories perfectly exemplify this. New technological artifacts had been constructed using the up-to-date knowledge of physical laws to solve specific concrete problems, almost in a trial-and-error basis to create know-how-type of knowledge pushed by the needs of the industrial revolution. At some point, these concrete artifacts were conceptualized as abstract objects toward a scientific theory with its own methods, proofs, and research questions, constituting a relatively autonomous science of specific technological objects. The new established science not only indicates how to improve the efficiency of existing techniques and/or artifacts but also (and very importantly) defines their fundamental characteristics, conditions, and limits.

Example 1.4 *Information theory.* Although large-scale communications systems had already been deployed for some decades before the 1940s, the engineers considered that errors in transmission were somehow inevitable. This commonsense practical knowledge was scientifically proven false when Claude Shannon published *A Mathematical Theory of Communication* [12] formulating the concept of *information entropy* and *mutual information*. Using these concepts, Shannon mathematically proved the existence of a code that leads to error-free communication if, and only if, the coding rate is below the channel capacity. This theory proposed in 1948 opened up a new field of theoretical research and also oriented practical deployments by giving an absolute indication of how far from the fundamental limit specific technologies are. It is noteworthy that, although Shannon had mathematically proven the existence of capacity-achieving codes, he has not indicated how to practically design them. For many years, researchers and engineers have pushed the technological boundaries and have developed different coding schemes. Only with the new millennium, feasible solutions have been proposed (or rediscovered) and, currently, the turbo codes and low-density parity-check (LDPC) codes are feasible options to reach a performance close to Shannon's limit. These high-performance techniques are used for example in cellular networks and satellite communications. The fundamental limit proposed by Shannon, though, cannot be surpassed by any existing or future technologies. A similar development happened in physics when the fundamental laws and limits of thermodynamics; firstly motivated by the development of thermal engines, the thermodynamic laws imposed fundamental limits of all existing or future engines [10].

An important remark is that sciences as theoretical discourses are historical and objective, holding a truth value relative to what is scientifically known at that time considering limitations in both theoretical and experimental domains. In this sense, scientific practice is an open-ending activity constituted by historically established norms. These norms, which are not the same for the different sciences and are internally defined through the scientific practice, determine the valid methodologies to produce scientific knowledge. Once established, this knowledge can then be used as raw material not only for the scientific practice from where it originates but also it can be (directly or indirectly) employed by other practices. As demonstrated in, for example, Noble [13], Feenberg [14], the scientific and technical development as a historical phenomenon cannot be studied isolated from the society and its articulation with the social whole becomes necessary.

From this perspective, this book will pedagogically construct a scientific foundation for CPSs based on existing scientific concepts and theories without distorting and displacing their specific objects. The resulting general theory will then be used to explain and explore different particular existing realizations of the well-defined abstract scientific object called *CPS* following the three proposed scientific rationalities. We are now ready to discuss the book structure and its rationale to then start our theoretical tour.

1.5 Book Structure

This book is divided into three main parts with ten core chapters, plus this introduction and the last chapter with my final words. The first part covers Chapters from 2 to 6, and focuses on the key concepts and theories required to propose a new theory for CPSs, which is presented in Chapters 7 and 8 (the second part of this book). The third part (Chapters 9, 10, and 11) deals with existing enabling technologies, specific CPSs, and their social implications

Part 1 starts with *systems* – the focus of Chapter 2, where we will revisit the basis of system engineering and then propose a way to demarcate particular systems following a cybernetic approach. Chapter 3 focuses on how to quantify *uncertainty* by reviewing the basis of probability theory and the concept of random variable. In Chapter 4, we will first define the concept *information* based on uncertainty resolution and then discuss its different key aspects, which includes the relation between data and information, as well as its fundamental limits. Chapter 5 introduces the mathematical theory of graphs, which is applied to scientifically understand interactions that form a *network* structure, from epidemiological processes to propagation of fake news. *Decisions* that determine *actions* are the theme of Chapter 6 discussing different forms of decision-making processes based on uncertainty, networks, and availability of information. Since decisions are generally associated with actions, *agents* are also introduced, serving as a transition to the second part.

Part 2 is composed of two dense chapters. Chapter 7 introduces the concept of *CPS as constituted by three layers*, which are interrelated and lead to a self-developing system. In Chapter 8, such a characteristic is further explored by introducing different approaches to model the *dynamics of CPSs*, also indicating performance metrics and their possible optimization, as well as vulnerabilities to different kinds of attacks. With these scientific abstractions, we will be equipped to assess existing technologies and their potential effects, which is the focus of the third part of this book.

Part 3 then covers concrete technologies and their impacts. Chapter 9 presents the *key enabling ICTs* that are necessary for the promising widespread of CPSs. Chapter 10 aims at different real-world applications that, following our theory, are conceptualized as realizations of CPSs. Chapter 11 is devoted to aspects *beyond technology* related to governance models, social implications, and military use.

At the end of each chapter, a summary of the key concepts accompanied by the most relevant references are presented followed by exercises that are proposed for the readers to actively learn how to operate with the main concepts.

1.6 Summary

In this chapter, we highlighted the reasons why a general scientific theory of CPSs is needed. We have briefly reviewed the beginnings of two related scientific fields, namely control theory [4] and information theory [12], contrasting them with the more philosophically leaned cybernetics as introduced by Wiener in [6]. To avoid potential threats of theoretical displacements of scientific theories, we have explicitly stated the philosophical standpoint taken in this book: science is a formal discourse holding true objective knowledge about well-defined abstract objects, which produces a knowledge effect on particular concrete objects. Scientific theories are then the result of a theoretical practice that produces new knowledge from historically determined facts and knowledges following a historically determined normative method of derivation and/or verification, which depends on the science/object under consideration. This leads to a philosophical classification that identifies three broad classes of scientific rationalities, helping to avoid misunderstandings of scientific results. The philosophical position taken here follows the key insights introduced by L. Althusser [8, 9], and I. Prigogine and I. Stengers [10]; the classification of different scientific rationalities is motivated by the work of Lepskiy et al. [11, 15] (although I do not share their philosophical position).

Exercises

1.1 **Daily language and scientific concepts.** The idea is to think about the word *power*.
 (a) Check in the dictionary the meaning of the word *power* and write down its different meanings.
 (b) Compare (a) with the meaning of the concept as defined in physics: ***power** is the amount of energy transferred or converted per unit time.*
 (c) Write one paragraph indicating how the daily language defined in (a) may affect the practice of a scientist working with the physical concept indicated in (b).

1.2 **Scientific rationalities.** In Example 1.3 in Section 1.4, an example of the three scientific rationalities was presented. It is your turn to follow the same steps.
 (a) Find an example of scientific practices that can be related to the three rationalities.
 (b) Based on (a), provide one example of each scientific rationality: (i) *classical*, (ii) *nonclassical*, and (iii) *interventionist*.
 (c) Articulate the scientific practices defined in (a) and (b) with other practices present in the social whole.

1.3 Alan Turing and theoretical computer sciences. The seminal work of Claude Shannon was presented in this chapter as the beginning of information theory. Alan Turing, a British mathematician, is also a well-known scientist considered by many as the father of theoretical computer sciences. Let us investigate him.

(a) Read the entry "Alan Turing" from The Stanford Encyclopedia of Philosophy [16].

(b) Establish the historical background of Turing's seminal work *On computable numbers, with an application to the Entscheidungsproblem* [17] following Example 1.4 in Section 1.4. This is the first paragraph of the text: *The 'computable' numbers may be described briefly as the real numbers whose expressions as a decimal are calculable by finite means. Although the subject of this paper is ostensibly the computable numbers, it is almost equally easy to define and investigate computable functions of an integral variable or a real or computable variable, computable predicates, and so forth. The fundamental problems involved are, however, the same in each case, and I have chosen the computable numbers for explicit treatment as involving the least cumbrous technique. I hope shortly to give an account of the relations of the computable numbers, functions, and so forth to one another. This will include a development of the theory of functions of a real variable expressed in terms of computable numbers. According to my definition, a number is computable if its decimal can be written down by a machine.*

(c) Discuss the relation between the work presented in (b) with the more philosophically leaned (speculative) *Computing Machinery and Intelligence* [18]. This is the first paragraph of the text: *I propose to consider the question, "Can machines think?" This should begin with definitions of the meaning of the terms "machine" and "think". The definitions might be framed so as to reflect so far as possible the normal use of the words, but this attitude is dangerous. If the meaning of the words "machine" and "think" are to be found by examining how they are commonly used it is difficult to escape the conclusion that the meaning and the answer to the question, "Can machines think?" is to be sought in a statistical survey such as a Gallup poll. But this is absurd. Instead of attempting such a definition I shall replace the question by another, which is closely related to it and is expressed in relatively unambiguous words.*

References

1 National Institute of Standards and Technology. Cyber-physicawl systems; 2020. Last accessed 2 October 2020. https://www.nist.gov/el/cyber-physical-systems.

References | 17

2 Nardelli PHJ, Kühnlenz F. Why smart appliances may result in a stupid grid: examining the layers of the sociotechnical systems. IEEE Systems, Man, and Cybernetics Magazine. 2018;4(4):21–27.

3 Lewis FL. Applied Optimal Control and Estimation. Prentice Hall PTR; 1992.

4 Maxwell JC. On governors. Proceedings of the Royal Society of London. 1868;16(16):270–283. 10.1098/rspl.1867.0055.

5 Huurdeman AA. The Worldwide History of Telecommunications. John Wiley & Sons; 2003.

6 Wiener N. Cybernetics or Control and Communication in the Animal and the Machine. MIT press; 2019.

7 Gerovitch S. From Newspeak to Cyberspeak: A History of Soviet Cybernetics. MIT Press; 2004.

8 Althusser L. Philosophy for Non-Philosophers. Bloomsbury Publishing; 2017.

9 Althusser L. Philosophy and the Spontaneous Philosophy of the Scientists. Verso; 2012.

10 Prigogine I, Stengers I. Order Out of Chaos: Man's New Dialogue with Nature. Verso Books; 2018.

11 Lepskiy V. Evolution of Cybernetics: Philosophical and Methodological Analysis. Kybernetes; 2018.

12 Shannon CE. A mathematical theory of communication. The Bell System Technical Journal. 1948;27(3):379–423.

13 Noble DF. Forces of Production: A Social History of Industrial Automation. Routledge; 2017.

14 Feenberg A. Transforming Technology: A Critical Theory Revisited. Oxford University Press; 2002.

15 Umpleby SA, Medvedeva TA, Lepskiy V. Recent developments in cybernetics, from cognition to social systems. Cybernetics and Systems. 2019;50(4): 367–382.

16 Hodges A., Zalta EN, editor. Alan Turing. Metaphysics Research Lab, Stanford University; 2019. Last accessed 20 October 2020. https://plato.stanford.edu/archives/win2019/entries/turing/.

17 Turing AM. On computable numbers, with an application to the Entscheidungsproblem. Proceedings of the London Mathematical Society. 1937;2(1):230–265.

18 Turing AM. Computing machinery and intelligence. In: Parsing the Turing Test. Robert E., Gary R., Grace B., editors. Springer, Dordrecht; 2009. pp. 23–65.

Part I

2

System

This first part of the book focuses on the main theoretical concepts that are the necessary raw material to construct the theory of cyber-physical systems to be proposed in the second part of the book. In this chapter, we begin this journey with the concept of *system*. The idea is to review the different meanings of the word, starting from the dictionary definition and moving toward a more technical one, which is used in the discipline of *Systems Engineering*. From there, a general method to organically demarcate the boundaries of a particular functioning system and everything else (i.e. its environment) will be proposed. We will further postulate the conditions of the existence of that particular functioning system as such. These conditions are divided into three levels that articulate the relation between that particular system and its environment. Different ways to classify systems will also be presented followed by an initial analysis of *Maxwell's demon* – a well-known thought experiment that was proposed to challenge the second law of thermodynamics. With this chapter, we aim to clarify the relation between the scientific domain whose object is a functioning system in general and its possible particular material realizations. In this sense, we argue that this theoretical methodology is essential not only to scientifically understand systems in general but also to engineer new or rectify particular existing systems.

2.1 Introduction

Let me start by showing a dialogue I had with my 5-year-old daughter:

- What is a car?
- It is a system designed to take people from one place to another, in a faster way and with less effort than walking.
- *What is a system?*

Cyber-physical Systems: Theory, Methodology, and Applications, First Edition. Pedro H. J. Nardelli.

- Well (…), in this case, it is a machine composed of things working together to perform some action.
- *And, in the other cases?*

My goal in this brief section is to provide satisfactory answers to these two questions in italics. The first step is to check which are the definitions of "system" that the dictionary gives.

Definition 2.1 *System in [1]* (1) A regularly interacting or interdependent group of items forming a unified whole; (2) an organized set of doctrines, ideas, or principles usually intended to explain the arrangement or working of a systematic whole; (3a) an organized or established procedure; (3b) a manner of classifying, symbolizing, or schematizing; (4) harmonious arrangement or pattern; (5) an organized society or social situation regarded as stultifying or oppressive.

It is clear the meaning I used to answer to my daughter is (1). At this point, two remarks are needed. First, not everything that is a *regularly interacting or interdependent group of items forming a unified whole* constitutes a system. For example, a clock, a car, or a computer are usually considered systems, while open markets or football teams are not (even though they might). This relatively arbitrary daily usage of the word shall be abandoned later in this chapter. The second remark is that the different meanings of the same word may create a series of confusion. While the meaning (1) mostly refers to the concrete reality, (2) and (3) seem to refer to symbolic domains, (4) and (5) are fuzzier and more subjective. Since this book is about *engineering*, (1) is the most appropriate meaning as our starting point.

With those warnings given, we are almost ready to transform the word "system" – our first raw material – into a concept that will become operational in the proposed theory of cyber-physical systems. Before we begin, though, it is important to describe how we will proceed. The first move is to provide a descriptive technical definition of what a system is following the well-established discipline *Systems Engineering* [2]. After this presentation, we can finally specify how it is possible to demarcate and articulate a particular functioning system with respect to its environment.

2.2 Systems Engineering

This section revisits the pedagogical exposition of the main technical terminology employed in *Systems Engineering* following the textbook *Systems Engineering and Analysis* [2]. The definition below states its answer to the question *what is a system?*

Definition 2.2 *System* A system is a set of *components* whose individual operating parts have specific *attributes* that are combined to form certain *relations* to perform one or few specific *functions*.

This definition is constructed upon four concepts:

1) **Components:** the elementary material parts – the building blocks – of the system.
2) **Attributes:** the properties of the components, or of the system as a whole.
3) **Relations:** the connections between components.
4) **Function(s):** what the system needs to accomplish.

It is quite straightforward to think in these terms. There are some specific pieces that are combined to form a whole that, if put together in such an organized form, can perform a predetermined task. Think about the do-it-yourself trend of building furniture, or even a house.

It is also important to point out that systems are always (directly or indirectly) related to dynamical processes. In other words, systems change. These changes usually refer to the *state* (situation) of the system at a specific point in time and in space. A series of changes in its state is called *behavior*. A *process* then refers to a sequence of behaviors. The *function* of the system is the outcome of a process or a series of processes.

It should be clear that the function of a particular system cannot be performed by any of its individual components alone, reminding us that *the system is more than the sum of its parts*. Roughly speaking, each component has its own role based on its attributes and interrelations. Depending on the role in the system, the different components can be classified into one of the following categories.

1) **Structural components:** The (quasi-)static parts of the system.
2) **Operating components:** The parts that perform the processing.
3) **Flow components:** Whatever (e.g. material, data or energy) is processed by the systems.

Structural components usually provide the support for the operating components to process the flow components. Before entering the system, the flow components are called *inputs*; the flow components that leave the system after processing are called *outputs*. The process of *transformation* from inputs to outputs that is performed inside the system always requires a motive force, either internal or external to it. Note, though, that not all systems have such a transformation process as their main function; there are systems whose function is transportation (i.e. flow of materials or data), or structural support for other systems to work (i.e. a highway, a bridge, or a railway). Nonetheless, all functioning systems (actively or passively) do work in a physical sense, and thus, energy is always converted in, by, or through systems. Systems may also function in a hierarchy: components can be systems in their own right, but, with respect to another broader system, they are simply subsystems. There also exist subsystems composed of other subsystems, and such a regression may go on as needed.

The following example illustrates all that has been discussed so far in this chapter.

Example 2.1 *Car as a system.* A car is composed of several different components from the three classes as, for example:

1) **Structural:** Chassis, windows, springs, and wheels.
2) **Operating:** Engine, brake, fuel pump, and radiator.
3) **Flow:** Electricity, diesel, air, and persons.

These and all other possible components of the car together are combined and designed to function as a whole system in order to perform a specific *function*: transporting persons from one place to another. A car may be either in a moving *state* or in a static *state*. The *relation* between the *attributes* of two components of the car, engine (attribute: converting thermal energy into kinetic energy) and wheels (attribute: using the kinetic energy converted by the engine to move the car), determines the *behavior* of the car over time and space. This relation is fundamental for the car to accomplish its *function*. The wheels and the engine are *subsystems* of the car. For example, the engine analyzed as a system functions by *processing* fuel (its *input*) in order to *transform* it into heat that will be converted into motion (its *output*). While the engine is a subsystem of the car, the car can also be analyzed as a subsystem of a transportation system.

What is of utmost importance for analyzing and engineering a system is to demarcate its boundaries, limits, and scope considering its particular function(s) while articulating it with its external world, its environment. In this sense, the focus needs to be on functioning systems, or systems that have potential to function. This topic will be the focus of the next section.

2.3 Demarcation of Specific Systems

The definition of a particular system is usually arbitrary or assumed as given. When someone talks about a car, anyone listening should have a very clear idea about that system. But, this might be a trick: is a car without an engine, or wheels, still a car? Similar kinds of discussions are very common in other domains as well. For example: should an oat-based drink be called oat milk?

In the legal domain, this definition is normative, usually as result of a deliberative process so that a definition can be imposed to indicate which is the correct word to name a given thing. The same happens with the word "car." Each country has its own legal definition of what is considered a "car" indicating who is eligible to drive, or what taxes are to be paid. However, as we have discussed in the previous chapter, scientific concepts are different from words used in other practices: the former exist in a specific theoretical discourse whose meaning is completely defined by the formal structure of the theory they are part of. The scientific meaning of the word "car" then needs to be very well-defined if it is to be considered a scientific concept. The question remains: how could such a theoretical concept be defined? The answer, once again, might be elusive: it depends on the scientific theory you are dealing with.

Since this chapter is about systems and how to scientifically define particular systems, our answer to that question will follow the solution we offer to the *demarcation problem*, where the system boundaries need to be defined with respect to everything else, which is referred to as *environment*. In other words, the problem is how to draw a line of demarcation between the system and the environment. This demarcation should not only indicate a functional relation but also a structural one. In this sense, the demarcation problem concerns both (i) how the system functions and (ii) how it is articulated with the environment so that its functioning can endure.

Definition 2.3 *Demarcation problem*   The demarcation problem refers to the determination of the boundaries that constitute *a particular system* as such in relation to everything else, i.e. the environment, as well as the specific articulations between the system and the environment.

This definition indicates the theoretical challenges ahead of us. Two remarks are important: (i) the solution of the demarcation problem determines at the same time what is internal to a particular system, and what is not, and (ii) from the proposed demarcation, the articulations and interactions between that particular system and everything else need to the determined, which also determines its level of autonomy. Clearly, other systems are also part of the environment, as well as other instances of reality that affect and are affected by such a particular system under investigation. Looking at the broader picture without specifying any particular system, we see a *whole* that is constituted by several systems that are related to each other in different manner and degree. In this case, the whole is called complex while articulated in dominance so that some relations may subordinate the others, affecting the system at different levels and by different instances. The case of the smart meter and the smart/stupid grid presented in Chapter 1 illustrates this idea.

The following proposition solves the demarcation problem stated in Definition 2.3.

Proposition 2.1 *System demarcation and its conditions of existence* A given **particular system (PS)** differentiates itself from everything else through a **peculiar function (PF)**; conversely, without this function, there is no such a system. The PF, at the same time, determines the existence of that given system as such while demarcating its boundaries and relations to the environment. There are three general groups of necessary conditions for a given particular system to exist as such. They are:

C1 **Conditions of production** refer to the (physical/material) possibility that PF can be produced.

C2 **Conditions of reproduction** refer to conditions internal to PS that guarantee the recurrence of its operation to perform its PF.

C3 **External conditions** are aspects outside the PS boundaries that directly or indirectly affect C2.

From this proposition, we can derive some interesting consequences. First, the system demarcation refers only to particular realizations of systems, not systems in general; on the other hand, this demarcation approach only makes sense if a theory of systems in general (as the one postulated here) exists. Second, a particular system whose PF is physically impossible cannot exist as such (e.g. a communication system designed to work outside Shannon's limit cannot exist); however, in some particular cases, it might exist as a thought experiment to challenge such an impossibility (this is the case of Maxwell's demon to be studied later in this chapter). Third, even if such a particular system is possible, it may not exist owing to conditions internal to its operation, and thus, the system cannot be established to perform its PF (e.g. a quantum personal computer is not yet possible because of operational instabilities related to quantum phenomena that the current technology cannot solve, or the lack of personnel capable of operating a given machine). Fourth, external events may also affect the system's conditions of reproduction (e.g. an earthquake that devastates the transportation system of a given city, or the lack of investment in education). Lastly, systems are not only affected by external events but also affect what is external to them (e.g. air pollution).

The starting point to demarcate a particular system is its technical description, i.e. its components that are combined to perform a given function. However, as we have discussed, these elements and their combinations are not enough to solve the demarcation problem following Definition 2.3: we have to describe what is needed for the system to function! To make this discussion less abstract, let us return to our "car" example.

Example 2.2 *Demarcating car as a system.* Let us consider a particular car that already exists. In this case, we have:

PS Car composed of its structural, operating, and flow components (as briefly described in Example 2.1).
PF Convert fuel into kinetic energy that is transferred to four wheels in order to transport persons.
C1 It is possible to convert fuel (e.g. gasoline) into kinetic energy that can be transferred to the wheels through a mechanical structure. The engine is what allows this conversion.
C2 A proper maintenance of the car and its components, availability of fuel, a person capable of driving...
C3 Technical training for car maintenance, highways and streets with good quality, adverse conditions such as economic crisis, and extreme weather events...

As expected, PS, PF, and C1 are very well defined (although the list of the example above should be longer to be complete), while C2 and C3 could be as long as needed (but never exhaustive). The engineer or analyst task is to select C2 and C3 so that the most relevant instances and factors are included depending on the context they are dealing with.

Let us provide another example, now a new wind turbine that does not yet exist but is being designed.

Example 2.3 *Demarcating a new model of a wind turbine as a system.* Let us consider a different kind of wind turbine that could be used in apartment buildings. Such a system does not yet exist as a material entity but engineers have a conceptual model about it.

PS (a) Structural components: metal tower, new type of blades, connection to the grid; (b) operating components: power electronic devices and an electric generator; and (c) flow components: wind and electricity.
PF Convert kinetic energy from wind into electric energy.
C1 It is physically possible to convert the kinetic energy from wind into electric energy.
C2 The electricity generated has to be synchronized with the grid in case of alternating current (AC) with the same frequency by using power electronic devices, proper maintenance of the components, existence of wind with enough kinetic energy to allow for the power conversion, existence of protective devices against risk situations (e.g. strong winds, or overcurrent)...

C3 Battery storage in the building, management of electricity demand, investment programs to support renewable energy, elections, subsidies, willingness to use wind turbines, regulations and laws that allow distributed energy resources, raw material to produce such a new turbine, extreme situations like a civil war...

As before, PS, PF, and C1 are very precise, while C2 and C3 are unbounded. In this case, this particular wind turbine does not materially exist, but it is still only a conceptual system. This nevertheless indicates the main determinations that a material realization of such a wind turbine are subjected to in order to function. The differences of material and conceptual systems will be discussed in the next section.

To conclude this section, we will present a proposition that posits the importance of the demarcation process.

Proposition 2.2 *Demarcation as a theoretical process* The demarcation of a particular system is a theoretical process that takes as raw materials the components of the system with their respective attributes and function(s) to determine its complex articulation with the environment for a given context. Therefore, the demarcation is a purely symbolic operation with respect to such a particular system, which becomes the abstracted, purified object of a discourse that aims to be objective and true. The demarcation process thus produces objective (scientific) knowledge about the actually functioning system through (i) its purification followed by (ii) its articulation in the specific context in which it functions, or will function.

It is also important to remember that the object of knowledge (the particular system that is subjected to the demarcation) is not the same as the material object (the real existing system), and they exist in different domains. However, the demarcation, as a theoretical process performed in the object of knowledge, produces a knowledge effect on the material object in the specific environment in which it exists or will potentially exist.

2.4 Classification of Systems

There are many ways in which systems can be classified according to their own characteristics or the focus of analysis. A proper classification is very important as it indicates the correct theoretical and experimental tools that are needed to develop objective knowledge of a particular system. In the following subsections, we will address a few general classes of systems, namely [2]: (i) natural and human-made, (ii) material and conceptual, (iii) static and dynamic, and (iv) closed and open. Other classes of systems will be defined in the upcoming chapters as we introduce new concepts.

2.4.1 Natural and Human-Made Systems

Through this classification, *natural systems* are the ones that come into being by natural processes without human intervention. Natural systems thus exist. *Human-made systems* are the ones that exist (or have potential to exist) by human intervention, i.e. humans are their agents of production. There is also a hybrid class, called *human-modified systems*, where either human made and natural subsystems are part of the same system, or humans directly intervene in natural systems.

The correct classification is clearly related to the demarcation of the system. In some sense, all systems on the planet Earth are human modified; however, classifying it as human made or natural may indicate the most important determinations for the particular system that is analyzed in the specific context in which it exists (or may potentially exist).

Example 2.4   The car and the wind turbine presented in the previous sections can be classified as human-made systems. The atmosphere of the planet Earth can be classified as a natural system as it has come into being without human intervention. However, in specific places where there are too many cars, or too many wind turbines, the atmosphere may change its internal dynamics because of different air composition that is due to chemical residues produced by cars or because of variations in air flows caused by turbines. In these cases, there is a human intervention and we could say that, in those specific regions, the atmosphere becomes a human-modified system.

2.4.2 Material and Conceptual Systems

Systems that manifest themselves in a material form are classified as *material systems*. They are composed of concrete components. Conceptual systems exist in a symbolic domain as plans, drawings, schemes, equations, specifications, or computer simulations used to create, produce, or improve material systems. Conceptual systems can also be concrete in some cases when a given material system is emulated on smaller scales or with elements of similar properties; in this case, a conceptual system is also a material system. There is also a possibility of hybrid material-conceptual systems as, for example, hardware-in-the-loop simulation platforms [3]. Further, the idea of digital twins can be seen as a hybrid system where there is a one-to-one map between an operating material system and its symbolic counterpart [4]. These last two approaches indicate some features of cyber-physical systems, but we are not yet ready to understand what *cyber* theoretically means.

Example 2.5 The wind turbine presented in Example 2.3 serves as an illustration of all cases. It is first a purely conceptual system where the components with their attributes are combined on paper based on dynamical equations, and it is then tested in a specialized computer simulator. The second phase is prototyping and a proof-of-concept phase on a small scale. This emulation of the material system to be produced is still conceptual as it is not implemented in its real conditions. The real wind turbine tested under controlled conditions, not yet connected to the real grid but to a virtual emulator building a hardware-in-the-loop simulation, is a hybrid conceptual-material system. The wind turbine in the field is a material system. If its operation is monitored and its operating behavior related to its digital twin, then we have another kind of hybrid system.

2.4.3 Static and Dynamic Systems

A static system is the one dominantly composed of structural components so that its state does not change, or changes in a negligible way, in time and space. A dynamic system is related to frequent changes in state; they are usually related to operating and flow components. Dynamic systems can have several subclasses, such as (i) linear or nonlinear, (ii) discrete time or continuous time, (iii) periodic or event-driven, (iv) deterministic or stochastic (or adaptive), (v) single input or multiple input, (vi) single output or multiple output, or (vii) stable or unstable (or chaotic). These characterizations are very important for any engineering system, either conceptual or material. The theory of dynamical systems – which is strongly mathematical – is at the core of most scientific and technological developments in the contemporary age [5]. New computer-based approaches are also becoming more and more prominent [6, 7], introducing new methods in different sciences and also in technical activities. Note that these methods refer to theoretical practices applied to conceptual systems, but that are used to implement and operate material systems. We will return to this in upcoming chapters where we will discuss artificial intelligence, self-organization, and agent-based models.

Example 2.6 The wind turbine of Example 2.3 once again provides us a good illustration. The tower of the wind turbine is a structural component (a subsystem) that is classified as static since it is not expected to change its state in time and space. The turbine itself (the rotating machine; an operating component) is a dynamical system whose changes are coupled with the wind movement (a flow component) that will determine the electric energy conversion, which can be measured by current and voltage as a function of time. It can be classified as a nonlinear continuous stochastic system.

2.4.4 Closed and Open Systems

Systems that have negligible interactions with their environments are called closed systems. In contrast, open systems are interwoven with their environment, exchanging data, matter, and energy. Despite all material systems being open, the concept of closed system is important to indicate the degree of interactions that are external to it. Moreover, such a concept has a great scientific value as it is used to define physical limits and laws of "purified" systems as an abstract object. Moving from a closed (abstracted) to a open (physical) system could be understood as a way to produce a material system from a conceptual one.

The demarcation of the system is key here as well. Once a particular system has its boundaries defined, it is possible to determine if it can be theoretically treated as a closed system. Usually, a closed system is associated with the analysis of the conditions of production, considering either the other two conditions of existence ideal for its functioning or completely neglecting them. The differentiation between closed and open systems is the basis for studying and quantifying their level of organization and uncertainty, as we will see in the next chapters.

Example 2.7 An experimental setting in a laboratory to test the wind turbine can be considered a closed system if everything needed to run such a test is contained there; there are no exchanges with the environment. A wind turbine in a real condition is an open system because it requires kinetic energy from the environment, it converts energy of another kind as an output to supply electricity to the environment, and also dissipates energy in the process of conversion.

2.5 Maxwell's Demon as a System

This section deals with an interesting problem of thermodynamics, a field of physics defined as follows [8]: *science of the relationship between heat, work, temperature, and energy. In broad terms, thermodynamics deals with the transfer of energy from one place to another and from one form to another. The key concept is that heat is a form of energy corresponding to a definite amount of mechanical work.* Among its fundamental laws, the first and second ones will be introduced here in brief. The first is the law of conservation of energy, which states that the change in the internal energy of a system is equal to the difference between the heat added to it from its respective environment and the work done by the system on its respective environment. The second law of thermodynamics asserts that entropy (which, in very rough terms, quantifies the degree of organization) of isolated (closed) systems (i) can never decrease over time, (ii) is constant if, and only if, all processes are reversible, and (iii) spontaneously tends to its maximum value, which is the thermodynamic equilibrium.

Such fundamental laws of physics were postulated in the nineteenth century when the study of heat transfer and heat engines was widespread because of the needs of the industrial revolution. As discussed in the previous chapter, this is another example of how relatively autonomous scientific knowledge can emerge from technical needs determined by a specific socioeconomic conjuncture [9]. This relative autonomy of the theory allows scientists to pose interesting thought experiments. One of the most famous is *Maxwell's demon*, in which the second law of thermodynamics would hypothetically be violated [10]. This experiment is defined next.

Definition 2.4 *Maxwell's demon [11]* *Maxwell's demon, hypothetical intelligent being (or a functionally equivalent device) capable of detecting and reacting to the motions of individual molecules. It was imagined by James Clerk Maxwell in 1871, to illustrate the possibility of violating the second law of thermodynamics. Essentially, this law states that heat does not naturally flow from a cool body to a warmer; work must be expended to make it do so. Maxwell envisioned two vessels containing gas at equal temperatures and joined by a small hole. The hole could be opened or closed at will by "a being" to allow individual molecules of gas to pass through. By passing only fast-moving molecules from vessel A to vessel B and only slow-moving ones from B to A, the demon would bring about an effective flow from A to B of molecular kinetic energy. This excess energy in B would be usable to perform work (e.g. by generating steam), and the system could be a working perpetual motion machine. By allowing all molecules to pass only from A to B, an even more readily useful difference in pressure would be created between the two vessels.*

Figure 2.1 depicts the situation. Our goal in this subsection is to analyze this problem as a system, indicating its (theoretical) conditions of existence and its PF. The idea here is not to solve this conundrum but rather analyze it as a system to properly define the problem and its characteristics. This should clear our path to theoretically work on the problem and then produce more knowledge about this object. Note that Maxwell's demon will also be studied in future chapters, and thus, it is worth for the reader to familiarize with it.

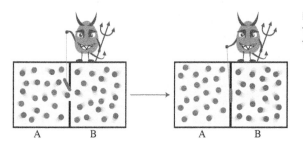

Figure 2.1 Illustration of the Maxwell's demon thought experiment.

2.5.1 System Demarcation

The Maxwell's demon thought experiment can be analyzed as a hypothetical system. The idea here is to provoke the reader to think about this procedure and be critical about it. Different from the other examples, we are dealing with something that does not and cannot exist in the real world like a car or a wind turbine. On the other hand, possible ways to realize this thought experiment have been presented in the literature, although this will not be our focus now. In the following, we propose the demarcation of the Maxwell's demon experiment as a system, indicating what this particular system, its PF, and its conditions of existence are as such.

PS (a) Structural components: a completely isolated box with a door that can open and close in an ideal way, (b) operating components: the demon who controls the door; and (c) flow components: a gas with equal temperature composed of molecules moving at different speeds.

PF Decrease the entropy of the system assuming no exchange of energy between it and its outside.

C1 It is physically possible to decrease the entropy of an isolated system (i.e. violating the second law of thermodynamics).

C2 The demon needs to know the velocity of the particles, their positions, and the sides that are associated with "cold" and "hot" states in order to control the door without using energy aiming at a decrease in the system entropy.

C3 The system has no relation to the environment (no flow of energy, matter, or information); therefore, this condition can be excluded.

2.5.2 Classification

The Maxwell's demon experiment is a human-made conceptual dynamic closed system. It is human made because this thought experiment only exists as a human-made theoretical construction. In this case, it is a conceptual system because it does not have a material realization. If we consider an experiment proposed in [12], then we would have a material system; this case will be analyzed later on. The system is dynamic because it changes its states over time. It is interesting to note that there are two interrelated levels with respect to the system dynamics: (i) system-level considering variations in macrostate properties like temperature or entropy, or (ii) molecular-level considering the movements of the molecules; these microstates are related to, for example, their individual velocity or position. The interrelation between the macrostates and the microstates is of extreme importance and will be presented in the next chapter, where we will focus on uncertainty. By definition, Maxwell's demon is an isolated system without any in- and outflows, and therefore, it is a closed system.

2.5.3 Discussions

Maxwell's demon is a theoretical construction, but it can be unambiguously defined as a system. The proposed demarcation indicates potential ways to actually build a material system that realizes the thought experiment. For example, the experiment presented in [12] defines a realization of the Maxwell's demon system based on electronics (single electron transistors). A careful analysis will show that such a material system maintains the basic features of the conceptual one, but with key differences that might be revealed by the demarcation of the actual experimental setup with its own limitations. In this case, the proposed demarcation is helpful either to design material realizations of a theoretical construction or to compare a physical experiment with the concept it aims to realize.

Another interesting point is about its PF and respective conditions of production. How is it possible to produce a function that does not respect a fundamental law of physics? A harsh answer would be that this system cannot exist, and thus, there is no need to discuss such a metaphysical construction in physics. We argue that the key issue here is the name given to the operating component: demon or a supernatural being. Its attribute is to open and close the door based on the knowledge of microstates of the system (velocity and location of molecules), and it has a specific goal of separating fast molecules to one side and slow to another – the hot and cold side, respectively. This split of the flow component based on its microstates into two different macrostates (hot and cold) leads to a decrease in the system entropy, which is the aim of the system. In today's terminology, Maxwell's demon could be renamed an *ideal smart controller*.

Clearly, with the knowledge and technology available in the nineteenth century, as well as the general context back then, defining the problem as a thought experiment in reference to a demon is understandable. The proposed change of name is an index that the demon is a computing device that is internal to the closed system. Since computing has its own fundamental laws related dissipation of energy as demonstrated by Landauer in [13], this result indicates that the Maxwell's demon experiment cannot form an isolated system because of the fundamental limits of computation. Therefore, the second law remains valid, and the Maxwell's demon experiment cannot exist as a system whose function is to decrease the entropy of an isolated system.

Given this, we can rectify our demarcation of Maxwell's demon as follows.

PS (a) Structural components: a completely isolated box with a door that can open and close in an ideal way, (b) operating components: an *ideal smart controller* which controls the door; and (c) flow components: a gas with equal temperature composed of molecules moving at different speeds.

PF Decrease the entropy of the system *through necessary computing processes* assuming no exchange of energy between it and its outside *except by the unavoidable dissipation related to computing processes.*

C1 It is physically possible to decrease the entropy of an isolated system *without violating the second law of thermodynamics by utilizing necessary computing processes.*

C2 The demon needs to know the velocity of the particles, their positions, and the sides that are associated with "cold" and "hot" states in order to control the door without using energy aiming at a decrease in the system entropy. *These are the computing processes.*

C3 The system has no relation to the environment (no flow of energy, matter, or information) *except by the fundamental heat generation related to the necessary computing processes.*

In this way, the conceptual system is posed in scientific terms that allow for experimentation by its material realization. As the result presented in [12] demonstrates, this thought experiment can be materially realized. However, we are not yet done with the Maxwell's demon experiment because of its relations to uncertainty, information, and decision-making: all topics related to the following chapters! The demon will stay with us for a while more.

2.6 Summary

In this chapter, we went through different meanings of the word "system." Among its different usages, we employed the definition and conceptualization from *Systems Engineering* as our theoretical raw material to then produce a scientific concept. From the general formalization of a system based on its components with their own specific attributes combined to perform a specific function or specific functions, we defined and solved the *demarcation problem*. The demarcation problem refers to how a particular system – determined by its peculiar function – is articulated with everything else by means of its conditions of existence. The demarcation is a theoretical process, exclusively symbolic, that produces objective knowledge of particular engineered objects that are conceptualized as functioning systems, which already exist or might exist. We also indicated different forms into which systems can be classified with respect to their own characteristics. Different examples were presented to illustrate the most important topics. At the end of this chapter, we dedicated a section to analyze as a system one classical thought experiment that will reappear in the following chapters: Maxwell's demon.

Exercises

2.1 Residential heating system. There are different ways to heat a house during cold periods as indicated by the USA Energy Department [14]. The idea here is to apply the concepts learned in this chapter to analyze residential heating.

(a) Consider an electric heating system connected to the main grid (as any other appliance of your house). Demarcate this system following Example 2.3.

(b) Classify the system demarcated in (a) following the examples presented in Section 2.4.

(c) During the winter months, the electricity demand in households with electric heating grows as the temperature decreases. Think about a heating system that could function without electricity from the grid. Demarcate this potential heating system and compare it with (a).

2.2 Boolean algebra and logic circuits. George Boole, an English mathematician and philosopher from the nineteenth century, proposed in his first book in 1847 a mathematical approach to logic by using mathematical symbols to represent classes of objects and then to manipulate them by mathematics [15]. In another groundbreaking work dating to 1938, Claude Shannon proposed in his Master's thesis a way to materially realize Boolean algebra by circuits [16]. They are represented by truth tables and logical circuits as illustrated in Figure 2.2.

(a) Analyze as a system (similar to Exercise 1) the logic gates AND, OR, and NOT, which are the basis of all logic circuits.

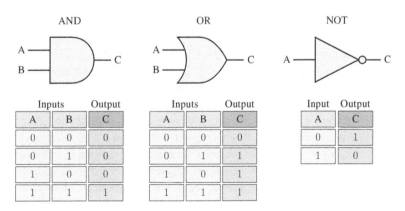

Figure 2.2 Truth tables and logic gates of AND, OR and NOT operations.

(b) Propose a way to materially realize these three logic gates based on (a).

(c) Follow the same steps used in (a) to analyze the proposal in (b), identifying the main differences between the conceptual system and its potential material realization.

(d) Read Shannon's Master's thesis [16] and discuss the importance of such a discovery.

PS The reader is suggested to play with online Boolean algebra calculators (e.g. [17]) as an extra task; a starting point could be the calculation of the truth tables presented in Figure 2.2.

References

1 Merriam-Webster Dictionary. System; 2020. Last accessed 22 October 2020. https://www.merriam-webster.com/dictionary/system.

2 Blanchard BS, Fabrycky WJ. Systems Engineering and Analysis: Pearson New International Edition. Pearson Higher Ed; 2013.

3 Isermann R, Schaffnit J, Sinsel S. Hardware-in-the-loop simulation for the design and testing of engine-control systems. Control Engineering Practice. 1999;7(5):643–653.

4 Haag S, Anderl R. Digital twin–proof of concept. Manufacturing Letters. 2018;15:64–66.

5 Beltrami E. Mathematics for Dynamic Modeling. Academic Press; 2014.

6 Wolfram S. A New Kind of Science. vol. 5. Wolfram Media, Champaign, IL; 2002.

7 Wolfram S. A class of models with the potential to represent fundamental physics. Complex Systems. 2020;29(2):107–147.

8 Drake GWF. Thermodynamics. Encyclopædia Britannica; 2020. Last accessed 10 November 2020. https://www.britannica.com/science/thermodynamics.

9 Prigogine I, Stengers I. Order out of chaos: Man's new dialogue with nature. Verso Books; 2018.

10 Thomson W. Kinetic Theory of the Dissipation of Energy. Nature. 1874;9:441–444.

11 Lotha G. The Editors of Encyclopaedia Britannica. Maxwell's demon. Encyclopædia Britannica; 2007. Last accessed 09 November 2020. https://www.britannica.com/science/Maxwells-demon.

12 Koski JV, Kutvonen A, Khaymovich IM, Ala-Nissila T, Pekola JP. On-chip Maxwell's demon as an information-powered refrigerator. Physical Review Letters. 2015;115(26):260602.

13 Landauer R. Irreversibility and heat generation in the computing process. IBM Journal of Research and Development. 1961;5(3):183–191.

14 USA Energy Department. Home Heating Systems. USA Energy Department; 2020. Last accessed 11 November 2020. https://www.energy.gov/energysaver/heat-and-cool/home-heating-systems.

15 Boole G. The Mathematical Analysis of Logic. Philosophical Library; 1847.

16 Shannon CE. A symbolic analysis of relay and switching circuits. Electrical Engineering. 1938;57(12):713–723. Available at: http://hdl.handle.net/1721.1/11173. Last accessed 11 November 2020.

17 WolframAlpha. Examples for Boolean Algebra. WolframAlpha; 2020. Last accessed 11 November 2020. https://www.wolframalpha.com/examples/mathematics/logic-and-set-theory/boolean-algebra/.

3

Uncertainty

This chapter focuses on *uncertainty* and the means to evaluate it through a mathematical theory of probability. Our objective here is to pedagogically present such a theory as a way to operate with random variables and random processes with special attention to well-defined systems following the demarcation process presented in Chapter 2. In this case, uncertainty can be related to either internal or external aspects of a given particular system, as well as to its articulation with its environment. We will introduce this theoretical field in an accessible manner with an explicit intention of reinforcing foundational concepts instead of detailing mathematical techniques and derivations. Such concepts are the raw material needed to understand the definition of *information* to be presented in Chapter 4. For readers with interest in the mathematics, the original book by Kolmogorov is a must [1]. Engineers and engineering students could also refer to [2] and readers with interest in complex systems to [3].

3.1 Introduction

The word "uncertainty" refers to aspects of a phenomenon, an object, a process, or a system that are totally or partly unknown, either in relative or absolute terms. For example, it is uncertain to me how many people will read this book, or if I will be able to travel to Brazil in the summer vacations. It is also uncertain to me who will win the lottery this evening; however, I am certain that I will not win because I have no ticket. Other phenomena involve even more fundamental types of uncertainty as, for instance, quantum mechanics (e.g. Heisenberg's uncertainty principle) or chaotic systems that are sensitive to initial conditions that are impossible to determine with the required infinite precision.

Probability theory is a branch of mathematics that formalizes the study of uncertainty through a mathematical characterization of random variables and random processes. Its basis is set theory, and it is considered to have been inaugurated as an

Cyber-physical Systems: Theory, Methodology, and Applications, First Edition. Pedro H. J. Nardelli.
© 2022 The Institute of Electrical and Electronics Engineers, Inc. Published 2022 by John Wiley & Sons, Inc.

autonomous field when Kolmogorov postulated the axioms of probability theory [1]. We can affirm that probability theory is a purely theoretical discipline related to purified mathematical objects. To avoid misunderstanding, it is worth mentioning that the treatment of uncertainty in real-world data is related to another discipline called statistics. While statistics is, in fact, founded on probability theory, it constitutes an autonomous technical discipline based on its own methodology and raw materials. Interestingly, the ongoing research efforts in machine learning and artificial intelligence are closely related to statistics and hence, probability theory. Some aspects of statistics will be discussed in forthcoming chapters.

Our focus here is only on the fundamentals of probability theory. We will mostly provide some intuitive concrete examples as illustrations, which are controlled experimental systems isolated from the environment (i.e. a closed system), and thus, a rigorous analysis of real-world data is put aside. In this sense, the most important concepts and the basics of their mathematical formulation will be stated in a comprehensive manner.

3.2 Games and Uncertainty

Most games involve some sort of randomness. Lottery, dices, urns, roulette, or even simple coin flipping are classical examples of games that are almost entirely a chance game. Those are usually one-shot games, and the uncertainty is directly related to the outcome in relation to choices the players have taken before the outcome is revealed. They are usual examples in probability theory textbooks.

Other classes of games, like chess or checkers, are strategic and, in some sense, deterministic. However, the (strategic) interactions between the two players in order to win also lead to uncertainty but of a different type in comparison with chance games. We can say that this is more an operational uncertainty in a sense that it is impossible to know with certainty how the process will evolve, except in very few (unrealistic) cases where very strong assumptions about the players are taken (e.g. both players will always act in the same way when faced with the same situation).

There are other kinds of games that combine both types of uncertainty. These are games where the initial state of the players is given by a drawing process (like a lottery), and the game itself is based on (strategic) decisions of players. Card games and dominoes are well-known examples. These are the most interesting games to exemplify and introduce the idea of random variables and random processes. Because of its relative simplicity, our discussions will mainly concern dominoes, whose main aspects will be defined next. Note that all games considered in this chapter are closed systems.

Definition 3.1 *Dominoes.* Dominoes is a tile-based game with 28 pieces forming a pair of two numbers ranging from 0 to 6, represented here as [·,·], without repeating pairs. For example, if the numbers in the tile are 4 and 3, we represent this as [4,3], or [3,4], because the pairs are unordered. Before starting, all tiles are placed with their numbers down and shuffled. The players (usually from two to four) take five pieces without the others seeing them and keep the leftovers as a bank. Whoever has the highest double (i.e. [6,6], [5,5], and so on) in the hands begins; if no one has a double, then the tile with the highest sum begins. With the first one in the deck, each player must pair off the numbers sequentially (e.g. clockwise). If a player in his/her turn does not have a tile with a number that matches the current deck, then he/she needs to get new pieces from the bank until he/she gets a suitable one. If the bank is empty, then that player misses the turn. The game ends when the first player has used all his/her pieces. If the game is blocked (i.e. no one has a piece to keep the game running), the winner is the one with the lowest sum in the hand.

A full set of dominoes is shown in Figure 3.1a, while a snapshot of a typical deck is presented in Figure 3.1b. This game is interesting because it provides a pedagogical way to present probability theory, ranging from the simplest case of one random variable to the most challenging cases of random processes. In more precise terms, there are two main sources of uncertainty in dominoes, namely (i) the drawing processes at the beginning and during the game, and (ii) the behavior of the players, who are dynamically interacting in a sequential process. In this case,

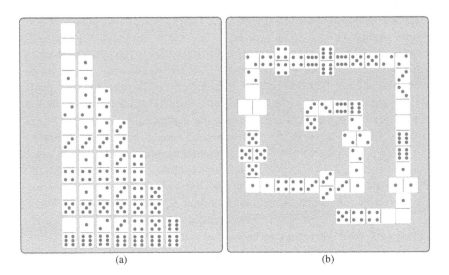

(a)   (b)

Figure 3.1 (a) Dominoes set; (b) example of a game.

interesting questions could be posed: (a) how frequently will a player begin with the same hand; (b) how frequently will all players begin with the same hand; and (c) how frequently will a game with all players starting with the same hand lead to the same playing sequence resulting in identical games? Without any formalism, even a child can readily answer that (a) is more frequent than (b) and (c), and (b) more frequent than (c). Our task is then not only to prove these inequalities but also to find a way to quantify the chances of these outcomes. Let us move step-by-step in a series of examples.

Example 3.1 *Drawing one specific tile.* Consider that a given player takes only one tile from the full dominoes set without seeing its respective number. Before looking at it, any tile out of the 28 could be selected. In this case, we can say that the player has a chance of 1 out of 28 possible outcomes, which we denote by 1 : 28, to get any specific tile. Note that 1 : 28 is not the same as 1/28 (which means 1 divided by 28). Hence, this number 1 : 28 could be used to quantify the uncertainty of the outcome. After the player sees the content of the tile, the outcome is certain, and thus, the tile is exactly the selected one. For example, the chance of having the tile [6,6] is 1 : 28 *before* the selection is revealed. *After* that, there is no uncertainty: the selected tile is either [6,6] or not. Without the [6,6] in the set, the chance of any other remaining tile to be drawn is 1 out of 27 possible outcomes (1 : 27). If the [6,6] is returned to the set, the chance is again 1 : 28.

Example 3.2 *Drawing two specific tiles in different ways.* The player takes two tiles at the same time, placing one on the right side and the other on the left. What is the chance that the tiles are [0,0] and [6,6]? There are two possible favorable outcomes: (A) the tile on the right is [0,0] and the one on the left is [6,6], or (B) the tile on the right is [6,6] and the one on the left is [0,0]. In this case, the chances are 2 out of (28×27) possible outcomes, or 1 : 378, where $28 \times 27 = 756$ refers to all possibilities of pairs. If positions of the tiles matter (i.e. the favorable outcome is [6,6] on the right and [0,0] on the left), only outcome (A) is favorable, and thus, the chances are smaller: 1-out-of-756. This is similar to a case of subsequent drawings where the first tile needs to be [6,6] and the second needs to be [0,0]. This results in 1 out of 756 possible outcomes. Note, though, that the chance of losing this game in the first draw is 27 out of 28 possible outcomes. If the draw is subsequent but the order does not matter, then we are in the first case: 2 out of 756 possible outcomes (or 1 : 378) with a chance of losing in the first draw equals 26 out of 28.

Example 3.3 *One player draws two times with the same five tiles.* Consider a game with two players where each one takes five tiles. We can determine the chances of one player getting the same hand, assuming that he/she is the first to take from the deck all five tiles. In this case, the player knows the five tiles of the first game, and therefore, we need to find the chance of having the same five tiles. The number of overall possibilities is $28 \times 27 \times 26 \times 25 \times 24 = 11{,}793{,}600$. The favorable outcomes are all possible arrangements of these five tiles, which can be computed as five options for the first, four for the second, and so on. This leads to $5 \times 4 \times 3 \times 2 \times 1 = 120$. Therefore, the chance of having the same hand twice is 1 out of 98 280 possibilities. Another way to understand this is that the chance for the first tile being in the desired set is $5:28$, the second is $4:27$, until the fifth is $1:24$, reaching the same result. We could follow similar steps to find the solution for the second player and also in the case of sequential one by one draw (this will be the focus of Exercise 3.1). Note that the definition of the favorable outcomes is more intricate in this case since the chance of the second player to get a given tile after the first player refers to the chance: (i) the latter not having selected such a tile in his/her turn, and (ii) the former having selected it.

Now, we can provide an outline of how to answer the three questions posed before these four examples. For the question (a) how frequently will a player begin with the same hand, we should proceed similar to Example 3.3 but indicating the order in which the pieces are taken from the deck (i.e. whether the player is the first one to take the five tiles, or if there is a different procedure for taking the tiles). The idea is to check the favorable outcomes (the specific five tiles) with respect to the number of pieces in the deck. The question (b) how frequently will all players begin with the same hand follows the same procedure, but each player has to be analyzed individually, and all players must have the same five tiles. However, it is easy to see that the situation described in (a) is completely covered by the situation in (b), which specifies even further the favorable outcome. We can say that the favorable outcome related to (a) is a subset of the one described in (b). For being more restrictive, the chances of (b) are smaller than (a).

The last question (c) how frequently will a game with all players starting with the same hand lead to the same playing sequence becoming identical games is trickier. First, the situation described in (b) is covered by the one described in (c), and thus, (c) has, at best, the same chance as (b). However, (c) involves an operational uncertainty related to how the game develops from the decisions and actions of the (strategic) players. Such decision-making processes are, in principle, unknown, and thus, we cannot compute the chances related to them without imposing (strong) assumptions. For example, if we assume that (i) all players will always take the same decisions and actions when they are in the same situation, and (ii) they play in the same sequence, then if they have the same tiles, the outcome of the game will be the same, reducing the chances of (c) happening to the outcome that all players have the same hands, which is the situation described in (b). In this case, the chances of (b) and (c) are the same. Nevertheless, any other assumption related to the players' behavior would imply more possibilities and therefore, the chances of (c) to occur would be lower than (b), and thus, (a). The situation described in (a) and (b) refers to *random variables*, while (c) refers to *random processes* or *stochastic processes*. Random variables are then associated with single observations, while stochastic processes with sequential observations constituted by indexed (e.g. time- or event-stamped) random variables considering the order of the observations. More details can be found in [3].

Besides, an attentive reader will probably not be satisfied because the questions talk about the *frequency* of a specific outcome and the answers displace the question of talking about the *chances* of those outcomes. This is a correct concern. The fundamental fact taken into account in those cases is that the studied outcomes are one-shot realizations of either a given process of drawing as in (a) and (b), or of a well-determined dominoes complete match as in (c). Since each one-shot realization of the process is unrelated to both previous and future realizations, the outcomes are then independent. A deeper discussion about the dependence and independence of outcomes will be provided later in this chapter. Let us now exemplify this by revisiting Example 3.1.

Example 3.4 *Chance and frequency of drawing one specific tile.* It was shown that the chance that a player gets a specific tile, let us say [6,6], from the full dominoes set is 1 out of 28 possible tiles. In other words, there is a chance of 1 : 28 of selecting [6,6] in a one-shot trial. If the tile is returned and the tiles are mixed again, the chance of getting [6,6] in the next one-shot trial is the same 1 : 28. If a new trial happens following the same procedure, the chances are the same 1 : 28. This could go indefinitely with the same chance of 1 : 28.

A skeptical person is concerned with the fairness of the game because one player gets [6,6] three times in a row. He/she then decides to take notes on how many times [6,6] appears as an outcome by filling a table with two possible outcomes **[6,6]** and **not [6,6]**. After 2,800 trials, the person decides to count the frequency of the outcomes. The **[6,6]** column is marked 102 times out of 2800 outcomes, leading to an approximate frequency of 1 : 28. The **not [6,6]** column is marked 2698 times out of 2800 outcomes, leading to an approximate frequency of 27 : 28. In this case, the frequency of outcomes is numerically approximated by the chance of the specific outcome in a one-shot trial. In addition, the sum of the events of interest, **[6,6]** and **not [6,6]**, results in 2800, leading to 2800 times out of 2800 outcomes, and therefore equals 28 : 28, or 1 : 1 (which is a certain outcome).

With this example, we have shown that the frequency of the outcomes in a (fair) drawing process can be approximated by the chance of a given outcome in a one-shot realization. If the drawing process is repeated many many times, then the frequency will tend to have the same value as the chance of a one-shot realization. For the questions asked in (a), (b), and (c), this fact is enough to sustain that frequency and chance have the same numerical value. But we need to go further, and this will require some formalization of the concepts presented so far in order to extend them. In the following section, we will start our quick travel across the probability theory world always trying to provide concrete examples.

3.3 Uncertainty and Probability Theory

Consider a particular system or process containing different observable attributes that can be unambiguously defined. For example, the dominoes set has 28 tiles with two numbers between zero and six, coins have two sides (heads and tails), the weather outside can be rainy or not, or the temperature outside my apartment during July could be any temperature between -10 and $35\,°C$. Given these observations, there is an unlimited number of possible *experiments* that could be proposed. One experiment could be related to measuring the temperature outside for the whole July, every day at 7 am and 7 pm. Another experiment could be to throw a coin five times in a row and then record the sequence of heads and tails. The examples provided with dominoes in the previous sections are all different experiments related to drawing tiles of the dominoes set.

Each experiment is thus defined by *events* related to observations of a set of predefined attributes of the system following a well-defined protocol of action (i.e. a procedure of how to observe and record observations). Depending on the specific case, the result of an observation is referred to as *outcome, sample, measurement*, or *realization*. If, before the observation, the attribute to be observed has its outcome already known with certainty, the system or process is called deterministic. If this is not the case, there is uncertainty involved, and thus, the possible set of outcomes is associated with *chance*, or *probability*.

The formal notation to be employed here will be presented next.

Definition 3.2 *Notations.*
- A particular system or process Φ is associated with $i \in \mathbb{N}^+$ different observable attributes $a_i \in \mathcal{A}_i$, where $\mathcal{A}_i$ is the set composed of all possible outcomes of the ith attribute.
- Each observation process is determined by a specific protocol ρ_i that unambiguously determines $\mathcal{A}_i$.
- The result of each observation $k_i \in \mathbb{N}$ of the attribute a_i is denoted $a_i[k_i]$, then $a_i[k_i] \in \mathcal{A}_i, \forall\, k_i$.
- An experiment Ξ is designed with respect to Φ by defining $j \in \mathbb{N}^+$ events of interest, denoted by $E_j \subseteq S_\Xi$, such that $\mu : \{\mathcal{A}_1, \ldots, \mathcal{A}_i\} \to S_\Xi, \forall j$, where $\mu(\cdot)$ is a function that maps the set of all possible observation outcomes $\mathcal{A}_i$ into the *sample space* S_Ξ.

- S_Ξ covers all possible outcomes of the experiment Ξ, including the possibility of $n + 1$ sequential observations $k_i, \ldots, k_i + n$ of the same attribute a_i with $n \in \mathbb{N}$.
- If the outcome of Ξ is known with certainty, the system or process Φ is said to be deterministic with respect to the experiment Ξ, and thus, the event E_j will either happen or not.
- If the outcome of Ξ is unknown but S_Ξ is known with certainty, then the system or process Φ is random with respect to Ξ; the chances that the event E_j will happen can be quantified by a function P named *probability measure*, which is defined in Definition 3.3.

It is noteworthy that the proposed notation formalizes the observation process with respect to a system or process. This is a remarkable difference from most textbooks, e.g. [2], where neither the observation protocol grounded in a particular system or process is formalized nor its randomness is formally defined with respect to the experiment under consideration. Although these aspects are found in most books in a practical state, our aim here is to explicitly formalize them in order to support the assessment and evaluation of the uncertainty of different systems and processes. To do so, we first need to mathematically define *probability measure*.

Definition 3.3 *Probability measure.* Any function P that satisfies the following three axioms is denominated as a probability measure over the sample space S.

- **(A1)** For any subset $E \subset S$, $P(E) \geq 0$.
- **(A2)** $P(S) = 1$.
- **(A3)** For disjoint sets $E_1 \subset S$ and $E_2 \subset S$ so that $E_1 \cap E_2 = \emptyset$, $P(E_1 \cup E_2) = P(E_1) + P(E_2)$.

These are the well-known axioms of probability proposed by Kolmogorov [1]. In plain words: probability is positive with a maximum value of 1, where the probability of two disjoint sets is the sum of their individual probabilities. With those three basic axioms, it is possible to derive several properties of P [1, 2].

With Definitions 3.2 and 3.3, we can return to the dominoes but now discussing probabilities.

Example 3.5 *Probability of drawing one specific tile.*
- System Φ: a dominoes set with 28 pieces.
- Protocol ρ: (i) all tiles are facing down, (ii) they are mixed, (iii) one tile is taken, (iv) its attribute a that will be defined next is observed and recorded, and (v) the tile is returned.
- Attributes $a \in \mathcal{A}$: the two numbers written in the tile so that $\mathcal{A} = \{[0,0], [0,1], \ldots, [6,6]\}$. Note that in the dominoes set, the two numbers form a unordered pair so that $[0,1]$ is equivalent to $[1,0]$ and so on.

Experiment 3.1

- Experiment Ξ_1: Perform the protocol ρ one time, leading to a sample space $S_{\Xi_1} = \mathcal{A}$.
- Event of interest E_1: Observe $a = [6, 6]$ as the outcome of the experiment Ξ_1.
- The outcome is uncertain because of the protocol ρ and the definition of Ξ_1, but the outcomes are fair (i.e. equally likely).
- If the protocol ρ leads to a fair selection where all tiles have the same likelihood to be taken, then the probability p_1 of taking any specific tile is the same.
- Since the tiles are mutually exclusive (once one is selected, all the others are not), and thus, form disjoint sets whose union is $\mathcal{A}$, we can apply (A3): $P(\mathcal{A}) = P(\text{the observed tile is } [0,0]) + \cdots + P(\text{the observed tile is } [6,6]) = 28p_1$
- Since $\mathcal{A} = S_{\Xi_1}$, then we can apply (A2) so that $P(\mathcal{A}) = P(S_{\Xi_1}) = 1$, leading to $p_1 = 1/28 = 0.0357 = 3.57\%$.
- Finally, $P(E_1) = p_1 = 1/28$, $P(\text{not } E_1) = P(S_{\Xi_1}) - P(E_1) = 1 - p_1 = 27/28 = 0.9643 = 96.43\%$.

Experiment 3.2

- Experiment Ξ_2: Perform the protocol ρ three times, leading to a sample space $S_{\Xi_2} = \mathcal{A} \times \mathcal{A} \times \mathcal{A}$, where the symbol "$\times$" denotes in this case the Cartesian product of sets.
- Event of interest E_2: Observe $\{a[0] = [6,6], a[1] = [6, 6], a[2] = [6, 6]\}$ as the outcome of the experiment Ξ_2.
- If the protocol ρ leads to a fair selection where all the tiles have the same likelihood to be taken, then the probability p_2 of taking any specific combination of three tiles is the same.
- Since the tiles are mutually exclusive, and thus, form disjoint sets whose union is $\mathcal{A}$, we can apply (A3): $P(\mathcal{A}) = P(\text{the observed tiles are } \{[0,0], [0,0], [0,0]\}) + \cdots + P(\text{the observed tiles are } \{[6,6], [6,6], [6,6]\}) = 28^3 p_2 = 21,952 p_2$
- Since $\mathcal{A} \times \mathcal{A} \times \mathcal{A} = S_{\Xi_2}$, then we can apply (A2) so that $P(\mathcal{A} \times \mathcal{A} \times \mathcal{A}) = P(S_{\Xi_2}) = 1$, leading to $p_2 = 1/21\,952 = 0.0046\%$.
- Finally, $P(E_2) = p_2 = 1/21\,952$, $P(\text{not } E_2) = P(S_{\Xi_2}) - P(E_2) = 1 - p_2 = 21\,951/21\,952 = 99.9954\%$.

What is interesting is that one can propose an unlimited number of experiments following the same basic observation process related to a particular system. In this case, it is also important to have a more detailed characterization of it, which can then be used to solve questions related to different experiments grounded in that process. One important tool is the frequency diagrams or histograms. The idea is quite simple: repeat the observation process several times, counting how many times each attribute is observed, and then plot it with a bar diagram associating

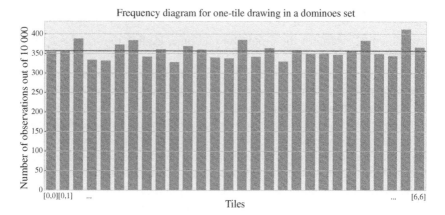

Figure 3.2 Example of a frequency diagram related to the observations of the attributes $a \in \mathcal{A}$ (two numbers in the tile) of the system Φ (a dominoes set composed of 28 tiles) following the observation protocol ρ repeated for 10 000 times. The dark gray line indicates the expected number of occurrences for each tile.

each attribute with its respective number of observations. Figure 3.2 illustrates this for the case of the dominoes set.

In this case, the dark gray line is the *expected number* of occurrences of each tile; this number is intuitively defined as the probability of a given outcome (in this case 1/28) times the number of observations (in this case 10 000). This leads to an expected number of 357.14 observations for each different tile. Let us try to better understand this with the following examples.

Example 3.6 *Number on the left, number on the right.* We are now interested in checking the frequency of the different numbers in a tile, the one on the left, the other on the right. Different from the previous cases, the order of the pair in the tile matters because of the two different orientation options according to which the tile could be placed. This implies a change in our basic protocol ρ. The new protocol ρ' and attribute set $\mathcal{A}'$ are defined next.

- **System Φ:** A dominoes set with 28 pieces.
- **Protocol ρ':** (i) All tiles are facing down, (ii) they are mixed, (iii) one tile is taken and horizontally positioned so that one number is on the left and the other on the right, (iv) its attribute a' that will be defined next is observed and recorded, and (v) the tile is returned.
- **Attributes $a' \in \mathcal{A}'$:** The two numbers written in the tile where one number is on the left, the other is on the right (see Figure 3.3). Formally, $a = [a_{\text{left}}, a_{\text{right}}]$ with $\mathcal{A}_{\text{left}} = \mathcal{A}_{\text{right}} = \{0, 1, \dots, 6\}$.

Figure 3.3   Domino tile in horizontal position: one number at the left side, another at the right side.

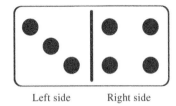

Left side   Right side

Note that $\mathcal{A}'$ has 49 elements but the dominoes have only 28 tiles. The difference comes from the observation protocols ρ and ρ'. In fact, the protocol ρ' brings a new uncertainty to the observation process: the position where the number will appear in the tile. We can analyze this situation by considering two different classes of tiles: doubles and not doubles. There are seven doubles where there is no uncertainty about the sequence in which the numbers will appear. The not-double class is composed of the other 21 tiles, where the sequence of appearance of the number does matter: the attribute [0,1] is different from [1,0], and so on. In this case, the sample space has a twofold increase, leading to 41 possibilities. The new sample space has then 41 elements. Using these numbers, we can plot the frequency diagram as presented in Figure 3.4. Since the protocol does not favor any specific outcome, the chance of having one specific value is 1 out of 7 possibilities (or simply 1 : 7) for both sides, leading to a probability of 1/7, because the outcomes are equally likely. If we repeat the observation process 10 000 times, the expected number of observed values is then $1/7 \times 10\,000$, which is equal to 1428.57, regardless of the specific number and the side.

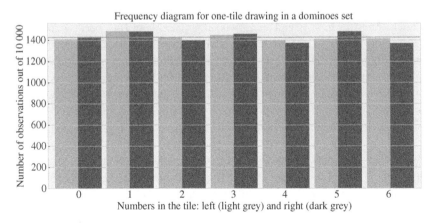

Figure 3.4   Example of a frequency diagram related to the observations of the attributes $a' = [a_{\text{left}}, a_{\text{right}}]$ with $\mathcal{A}_{\text{left}} = \mathcal{A}_{\text{right}} = \{0, 1, \ldots, 6\}$ of the system Φ (a dominoes set composed of 28 tiles) following the observation protocol ρ' repeated for 10 000 times.

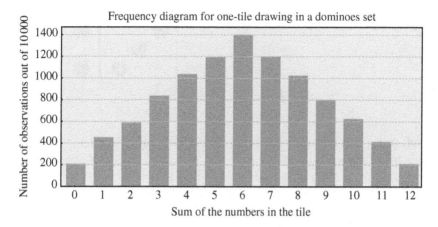

Figure 3.5 Example of a frequency diagram related to the attribute $a'' = a_{\text{left}} + a_{\text{right}}$ with $\mathcal{A}_{\text{sum}} = \{0, \dots, 12\}$ of the system Φ (dominoes set composed of 28 tiles) following the observation protocol ρ'' repeated for 10 000 times.

Example 3.7 *Sum of the two numbers in the tile.* Consider another observation protocol where we are interested in the sum of the number on the right and on the left. We can define the new attribute as $a'' = a_{\text{left}} + a_{\text{right}}$ with $\mathcal{A}_{\text{sum}} = \{0 + 0, \dots, 6 + 6\} = \{0, \dots, 12\}$. We have the new protocol ρ'' defined as: (i) all tiles are facing down, (ii) they are mixed, (iii) one tile is taken, (iv) its attribute a'' is observed and recorded, and (v) the tile is returned. Figure 3.5 illustrates the frequency diagram of this sum. The situation is now more challenging, because the expected number of observations of each attribute is not the same. By inspection, it is easy to see that $a'' = 0$ only if $a_{\text{left}} = 0$ and $a_{\text{right}} = 0$, which is $1:7$ and $1:7$, which results in $1:49$ and an associated probability of $1/49$, because the outcomes of a_{left} and a_{right} are equally likely. Considering the 10 000 observations, it is expected that $a'' = 0$ will be observed $1/49$ times 10 000, which is equal to 204.08. For $a'' = 1$, there are two possibilities: $[0,1]$ or $[1,0]$. In this case, there is twice the chance compared with the previous case, i.e. 2 out of 49 possibilities. Hence, the expected number of observations for $a'' = 1$ is 408.16. It is possible to follow a similar procedure for the other values of $a'' \in \mathcal{A}_{\text{sum}}$.

With these examples, we have shown how the different observation protocols define the observable outcomes and the respective sample spaces, which are the basis of specific experiments constructed to evaluate the uncertainty of particular systems or processes. Because the values of the attributes may vary at every observation, it is then possible to create a map that associates each possible observable attribute with a probability that it will be indeed observed as the outcome of one observation. Probability theory names a quantified version of this unknown observable attribute as *random variable* and the mathematical relation that maps the random variable with probabilities as *probability functions*; they are formally defined in the following.

Definition 3.4 *Random variable and probability functions.* Consider a system Φ associated with an observation process ρ and observable attributes $a \in \mathcal{A}$. A random variable $X(a)$ is a function that maps every attribute a with a real number so that $X : \mathcal{A} \to \mathbb{R}$.

[Discrete] A random variable is called discrete if its possible outcomes are defined in a countable set S_X, which can be either infinite, i.e. $S_X = \{x_1, x_2, \dots\}$, or finite, i.e. $S_X = \{x_1, x_2, \dots, x_n\}$ for a given number n. The *probability mass function* is defined for discrete random variables, i.e. $x \in S_X$, as:

$$p_X(x) = P(X = x) = P(\{a : X(a) = x\}), \tag{3.1}$$

where the function P satisfies the axioms defined in Definition 3.3.

[Continuous] A random variable is called continuous if its possible outcomes are defined in an uncountable set as, e.g. $\mathbb{R}$. The *cumulative density function* is then defined for $x \in \mathbb{R}$ as

$$F_X(x) = P(-\infty < X \leq x), \tag{3.2}$$

where the function P satisfies the axioms defined in Definition 3.3. The *probability density function* (if it exists) is then defined as:

$$f_X(x) = \frac{\mathrm{d}F_X(x)}{\mathrm{d}x}. \tag{3.3}$$

This can now be used not only to better characterize the uncertainty related to different processes and systems but also to develop a whole mathematical theory of probability. From the probability function, it is also possible to define a simpler characterization of random variables. For example, it is possible to indicate which is the outcome that is more frequent, or the average value of outcomes, or how much the outcomes vary. Among different possible measures, we will define next the *moments* of random variables, which can be directly employed to quantify their expected value and variance.

Definition 3.5 *Moments, expected value, and variance.* The kth moment of a given random variable X, $\mathbb{E}\left[X^k\right]$, is defined differently for discrete and continuous random variables as follows.

$$\mathbb{E}\left[X^k\right] = \sum_{x \in S_X} x^k \, p_X(x) \text{ for discrete } X, \tag{3.4}$$

$$\mathbb{E}\left[X^k\right] = \int_{-\infty}^{\infty} x^k \, f_X(x) \, \mathrm{d}x \text{ for continuous } X. \tag{3.5}$$

The expected or mean value of X is its first moment, and thus, $\mathbb{E}[X]$. The variance of X is a function of the first and second moments used to quantify the variation from the expected value and is computed as $var(X) = \mathbb{E}\left[X^2\right] - (\mathbb{E}[X])^2$.

Besides, it is worth introducing two other concepts that are helpful to analyze different observation processes. Consider a case of different observations of the same random variable X (i.e. all conditions of the observation process are the same). If a given observation of this process does not depend on past and future observations, we say that the realizations of X are independent and identically distributed (iid), which will lead to a memoryless process (a formal definition will be provided later). We can now illustrate these concepts with some examples.

Example 3.8 Consider the observation process defined in Example 3.5. We can create a discrete random variable $X(a)$ as:

- If $a = [0,0]$, then $X(a) = 1$,
- If $a = [1,0]$, then $X(a) = 2$,
- If $a = [2,0]$, then $X(a) = 3$,
- ...
- If $a = [6,6]$, then $X(a) = 28$.

Since the observation process leads to equally likely outcomes, the probability mass function is then $P(X = x) = 1/28$ for $x \in \{1, 2, \ldots, 28\}$. In this case, the different realizations of X are iid.

In this example, although it is possible to compute the expected value and the variance, these measures are meaningless because the mapping function $X(a)$ is arbitrary and its actual values are not related to the attribute a. Note that, in other cases, the mapping is straightforward from the measurement as, for instance, the number of persons in a queue. In summary, the correct approach to define the random variable under investigation depends on the system or process Φ, the observation protocol ρ, and the experiment Ξ of interest.

It is also important to mention that one interesting way to numerically estimate probability mass functions is by experiments like the ones presented in Figures 3.2, 3.4, and 3.5. Those results were obtained by computer simulations based on the Monte Carlo method [4]. The idea is to generate several observations and empirically find their probabilities as the number of times that a specific observation appears divided by the overall number of observations realized. Figure 3.6 presents analytical results generated from the probability mass function $P(X = x) = 1/28$ for $x \in \{1, 2, \ldots, 28\}$ and empirical results generated by computer simulations. Note that this random variable is *uniformly distributed* in relation to its sample space.

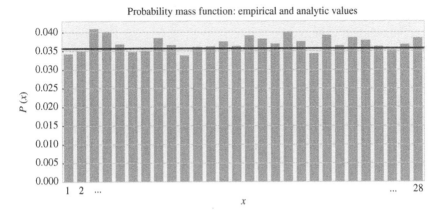

Figure 3.6 Example of a probability mass function diagram related to the random variable $X(a)$ obtained from the observations of the attributes $a \in \mathcal{A}$ (two numbers in the tile) of the system Φ (dominoes set composed of 28 tiles) following the observation protocol ρ repeated for 10 000 times. The bars are the empirical results, and the line indicates the analytic probabilities.

As a matter of fact, there are several already named probability distributions, for instance, Gaussian or Normal distribution, Poisson distribution, and Binomial distribution. Their mathematical characterization is available in different public repositories and textbooks, e.g. [2].

The importance of having a well-defined formulation of probability distributions is to perform computer experiments in a *generative* manner. The idea is to describe a real-world process or system with a known probability distribution and simulate its dynamics. Note also that, in many cases, the random variable values can be directly obtained from observations of the attributes, without defining a special mapping function as in Example 3.8.

Example 3.9 *Arrivals in a restaurant.* The manager of one restaurant is planning how many meals he/she should prepare every hour for the dinner. Without any good characterization, he/she estimated that from 5 pm to 6 pm there are usually fewer people than from 6 pm to 9 pm, which is the peak time. From 9 pm to 10 pm (when the restaurant closes), there is about the same number of people arriving as in the 5 pm to 6 pm period. He/she estimates the arrivals as follows.

- 5 pm to 6 pm and 9 pm to 10 pm: 5 persons per hour,
- 6 pm to 9 pm: 20 persons per hour.

However, he/she knows that the actual number may vary night after night, and that the numbers are somehow independent.

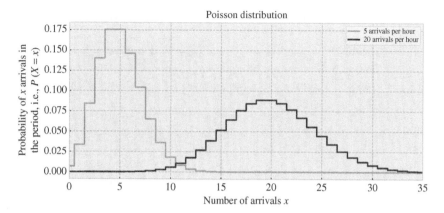

Figure 3.7 Poisson random variable considering two different client arrival rates in a period of one hour.

After joining a probability course, the manager discovered that this scenario could be studied by using Poisson random variables. The probability mass function is defined as:

$$P(x) = \frac{\lambda^x \exp(-\lambda)}{x!}, \tag{3.6}$$

where $\lambda > 0$ is the estimated number of arrivals per hour, $x \in \mathbb{N}$, and $x! = x \cdot (x-1) \cdot \ldots \cdot 2 \cdot 1$ is the factorial operation. For Poisson random variables, both the mean value $\mathbb{E}[X]$ and the variance $var(X)$ are equal to λ.

The manager then understands that the estimated arrival rate might be related to the mean value, and thus, he uses the aforementioned values as the parameter λ. Figure 3.7 shows the distributions defined by (3.6) considering $\lambda = 5$ and $\lambda = 20$. It is interesting to see that there is a great variation around the estimated number of arrivals, which was also indicated by the relatively high variance value. In the case of five arrivals per hour, there is a nonnegligible probability that 10 persons will arrive in that period, as well as none. However, the highest probability is associated with four or five arrivals. In this case, the probability of the random variable is $X = 4$ or $X = 5$ is 35%.

Now, looking at the case of 20 arrivals per hour, the situation is more challenging because the distribution is more spread, although the highest probabilities are $X = 19$ and $X = 20$, both below 10%. In plain words, this fatter distribution means that there is a relatively high probability that an outcome far from the estimated arrival rate is actually observed. This means that during one day between 7 pm and 8 pm, 30 persons may arrive, while on another day at the same time only 10 will come.

The probability mass function does not help the manager much to estimate how many meals he/she needs to prepare every evening, but it helps him/her to better understand the uncertainty involved in this case. Depending on other considerations, the manager could have other known distributions to model the problem. He/she could also acquire more fine-grained data and find an empirical distribution, possibly trying to match the empirical points with known distributions.

Example 3.10 *Maxwell's demon.* Consider the Maxwell's demon thought experiment introduced in the previous chapter. Let us consider two situations: (i) before the demon's interventions: the system is in a thermodynamic equilibrium, and thus, with the same temperature; and (ii) after the demon's interventions: each side is in a thermodynamic equilibrium but with different temperatures, one warmer, the other colder than before. From classical statistical mechanics, we know that the velocity of the particles for a given temperature is a random variable X that follows a Maxwell–Boltzmann probability distribution, whose probability density function is given by:

$$f_X(x) = \sqrt{\frac{2}{\pi}} \, \frac{x^2}{a^3} \exp\left(-\frac{x^2}{2a^2}\right), \tag{3.7}$$

where $x > 0$ and $a = \sqrt{k\frac{T}{m}}$ with T are the temperature, m the molecule mass, and k the Boltzmann constant.

Figure 3.7 presents the distribution (3.7) for $a = 1$ (arbitrary value) that characterizes how the velocity of the particles is distributed in the experiment for a given equilibrium temperature. It is interesting to see that the velocity of each molecule refers to the characteristic of one of the components of the system, which is composed of a very large number of molecules, while the temperature is a macrostate of the system (Figure 3.8).

The idea of the operation of Maxwell's demon is to separate the slow and fast molecules based on a given threshold (dashed line in the figure) to create a new equilibrium macrostate constituted by (a) slow molecules (cold side) and (b) fast molecules (hot side). In this case, Maxwell's demon is violating the second law of thermodynamics by creating a new equilibrium state (ii) where there will be two different probability distributions, one for (a) the other for (b). The formalization of this thought experiment and the estimation of the new equilibrium states will be the focus of Exercise 3.4.

Most of the examples presented in this section illustrate experiments where the different observations of the specific random variables are assumed iid. However, this is not always the case. Example 3.7 shows one type of dependence, because there is one new random variable defined as the sum of two other random variables. Another example is a drawing process presented in Example 3.1 but without returning the observed tile to the bank, which decreases the sample space, and thus, modifies the probability distribution. There are many other

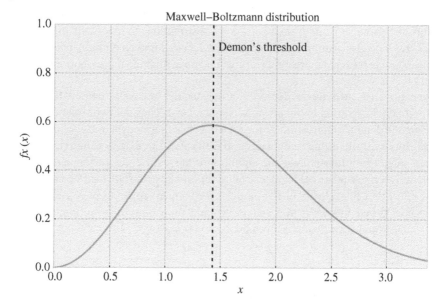

Figure 3.8 Maxwell–Boltzmann probability density function of the random variable that characterizes the speed of the gas molecules for $a = 1$ (arbitrary value) indicating the separation threshold used by Maxwell's demon to separate the fast and slow molecules.

modes of dependence between two or more random variables, which include operations and functions of random variables, and random processes. These topics will be the focus of our next section.

3.4 Random Variables: Dependence and Stochastic Processes

When thinking about systems and processes, we intuitively imagine that different observations of attributes and outcomes of experiments may be dependent. Observable attributes might be related to a background process or system, for example, when measuring temperature in a room using two sensors in different places, or the same sensor but whose measurements are taken every ten minutes. In both cases, if we consider that these observable attributes are associated with random variables, it is expected that they would be related to each other as they are measurements of the temperature (a physical property) of the same place. Hence, if the outcome of one of these random variables is known, the uncertainty of the other should decrease.

In other cases, the attributes might have another type of physical dependence, like the rotation speed of rotors in electric machines and the associated values of electric current and voltage. There is also the case of relations that are constructed in the design of the experiment of interest. For instance, in Example 3.7, the outcome of the experiment is constructed as the sum of two attributes that are characterized as two independent random variables.

Another type of dependence between random variables is related to sequential processes as exemplified by a dominoes game where the sample space dynamically changes at each action. A queue at the cashier of a supermarket also illustrates a sequential process where the number of persons waiting in the line to be served depends on the number of persons arriving in the queue and the service time that each person requires to leave the cashier.

To evaluate the dependence between random variables, it is necessary to formally define a few key concepts, as to be presented next.

Definition 3.6 *Relation between random variables.* Consider without loss of generality two random variables X and Y, with $x \in S_X$ and $y \in S_Y$, associated with the probabilities $P(\{X = x\})$ and $P(\{Y = y\})$. X and Y are called independent if their outcomes are unrelated; otherwise, X and Y are dependent, and this dependence may take different forms. For independent X and Y, we have:

- the joint probability $P(\{X = x\} \cap \{Y = y\}) = P(\{X = x\})P(\{Y = y\})$,
- the conditional probabilities $P(\{X = x\}|\{Y = y\}) = P(\{X = x\})$ and $P(\{Y = y\}|\{X = x\}) = P(\{Y = y\})$, where we read the probability of $X = x$ or $Y = y$ given that $Y = y$ or $X = x$, respectively.

If X and Y have the same sample spaces $S_X = S_Y$ and are associated with the same distribution function, they are then called *identical in distribution*. If X and $aY + b$ with $a, b \in \mathbb{R}$ are identical in distribution, then X and Y are of the *same type*.

To mathematically characterize the dependence/independence of random variables, we need to consider functions of two or more random variables. They are the joint, conditional, and marginal probabilities distributions. Therefrom, it is possible to propose a classification of types of uncertainties associated with different systems, processes, and experiments. Their formal definition is provided next.

Definition 3.7 *Joint, conditional, and marginal probability distributions.* Consider without loss of generality two random variables X and Y, with $x \in S_X$ and $y \in S_Y$.

[Discrete] The *joint probability mass function* is defined for the vector of discrete random variables $(x, y) \in S_X \times S_Y$, as:

$$p_{X,Y}(x, y) = P(\{X = x\} \cap \{Y = y\}) = P(X = x, Y = y). \tag{3.8}$$

where the function P satisfies the axioms defined in Definition 3.3.

The *marginal probability mass functions* are defined as:

$$p_X(x) = P(X = x, Y \in S_Y), \tag{3.9}$$

$$p_Y(y) = P(X \in S_X, Y = y), \tag{3.10}$$

from where the probability mass function of X and Y can respectively be obtained from the joint function by covering the whole domain of X and Y (i.e. the sample space S_X and S_Y).

The *conditional probability mass functions* are defined as:

$$p_{X|Y}(x|y) = P(X = x|Y = y) \tag{3.11}$$

$$p_{Y|X}(x|y) = P(Y = y|X = x) \tag{3.12}$$

where the symbol $X|Y$ reads "X given Y", and $Y|X$ reads "Y given X".

They are related through the following equality:

$$p_{X,Y}(x, y) = p_{X|Y}(x|y)\, p_Y(y) = p_{Y|X}(y|x)\, p_X(x), \tag{3.13}$$

which is the well-known Bayes' theorem.

[Continuous] The *joint cumulative density function* is then defined for $(x, y) \in \mathbb{R}^2$ as

$$F_{X,Y}(x, y) = P(\{X \leq x\} \cap \{Y \leq y\}) = P(X \leq x, Y \leq y), \tag{3.14}$$

where the function P satisfies the axioms defined in Definition 3.3.

The *joint probability density function* (if it exists) is then defined as:

$$f_{X,Y}(x, y) = \frac{\partial^2 F_{X,Y}(x, y)}{\partial x\, \partial y}. \tag{3.15}$$

The *marginal probability density function* of X in relation to Y is defined as:

$$f_X(x) = \int_{-\infty}^{\infty} f_{X,Y}(x, y')\, dy' \cdot f_Y(y) = \int_{-\infty}^{\infty} f_{X,Y}(x', y)\, dx'. \tag{3.16}$$

The *conditional probability density functions* are obtained by the following relation:

$$f_{X,Y}(x, y) = f_{X|Y}(x|y)\, f_Y(y) = f_{Y|X}(y|x)\, f_X(x). \tag{3.17}$$

Despite the potentially heavy mathematical notation, the concepts represented by those definitions are quite straightforward. The following example illustrates the key ideas.

Example 3.11 *Dominoes tile.* Consider the situation described in Example 3.6, where the observable attribute is $a' = [a_{\text{left}}, a_{\text{right}}]$ with $A_{\text{left}} = A_{\text{right}} = \{0, 1, \dots, 6\}$ following the protocol ρ'. We can directly associate a_{left} with a discrete random variable X and a_{right} with a discrete random variable Y, associated with the sample spaces $S_X = S_Y = \{0, 1, \dots, 6\}$. Because of the protocol ρ', the events are independent, and therefore, the outcomes of X and Y do not affect each other. We can then write:

- **Marginal:** $p_X(x) = P(\{X = x\}) = 1/7$ and $p_Y(y) = P(\{Y = y\}) = 1/7$,
- **Joint:** $p_{X,Y}(x, y) = P(\{X = x\} \cap \{Y = y\}) = P(X = x)P(Y = y) = (1/7)(1/7) = 1/49$ with the sampling space being $S_X \times S_Y = \{[0,0], \dots, [6,6]\}$, where the pairs are ordered (i.e. [0,1] is not equivalent to [1,0], and so on).
- **Conditional:** $p_{X|Y}(x|y) = p_X(x) = 1/7$ and $p_{Y|X}(y|x) = p_Y(y) = 1/7$.

We propose a slight change in the protocol ρ' so that the smaller number of the tile is always on the left. In this case, $a_{\text{left}} \leq a_{\text{right}}$, and thus, $X \leq Y$. For example, if it is known that $X = 6$, we are sure that $Y = 6$ as well. If $X = 5$, then there is still uncertainty since there are two possible values that Y can assume. The conditional probability mass function will then be $p_{Y|X}(y|x) = 1/(7 - x)$. If we consider the case that Y is known, then if we have $Y = 0$, then $X = 0$. If $Y = 1$, then X can assume two values, and so on. The conditional probability will be then $p_{X|Y}(x|y) = 1/(y + 1)$.

The joint probability is considerably simple: $p_{X,Y}(x, y) = 1/28$. This is the case because there is only one tile of each, and the difference is how to arrange them always keeping the smaller number on the left. With these distributions, we can easily find the marginal distributions as follows:

- $p_X(x) = \dfrac{p_{X,Y}(x, y)}{p_{Y|X}(y|x)} = \dfrac{7 - x}{28}$,

- $p_Y(y) = \dfrac{p_{X,Y}(x, y)}{p_{X|Y}(x|y)} = \dfrac{y + 1}{28}$.

Another possible way to create a dependence is by defining experiments where new random variables are defined as functions of already defined ones. For example, one new random variable can be a sum or a product of two other random variables. Example 3.7 intuitively introduces this idea. Despite the importance of this topic, the mathematical formulation is rather complicated, going beyond the introductory nature of this chapter. The reader can learn more about those topics, for example, in [2]. What is extremely important to remember is that one shall not manipulate random variables by usual algebraic manipulation; random variables must be manipulated according to probability theory.

Before closing this chapter, there is still one last topic: stochastic processes. A stochastic process, also called random process, is defined as a collection of random variables that are uniquely indexed by another set. For instance, this index set could be related to a timestamp. If we consider that the temperature measured by a given thermometer is a random variable, the index set can be a specific timestamp (i.e. year-day-hour-minute-second) of the measurement. The stochastic process is then a collection of single measurements – which is the random variable – indexed by the timestamp. A formal definition is presented next.

Definition 3.8 *Stochastic or random process.* Without loss of generality, consider:

- A particular system or process Φ is associated with one observable attribute $a \in \mathcal{A}$, where $\mathcal{A}$ is a set composed of all possible outcomes.
- Each observation process is determined by a specific protocol ρ that unambiguously determines $\mathcal{A}$.
- The result of each observation $k \in \mathbb{N}$ of the attribute a is denoted $a[k]$, then $a[k] \in \mathcal{A}$ where $k \in \mathcal{I}$.
- An experiment Ξ can be defined by associating $a[k]$ with a random variable $X(a[k])$, or simply $X(k)$, over the sample space $S_\Xi(k)$, which may vary for each k different observation.
- We can define a collection $\{X(0), X(1), \ldots X(k_{\max})\}$, where $\mathcal{I} = \{0,1,\ldots,k_{\max}\}$; this collection is named as a *stochastic* or *random process* with an index set $\mathcal{I}$.

Example 3.12 *Queue in the university restaurant.* The manager of the university restaurant is monitoring every half an hour, from 10:00 to 14:30, how many people are lined up in the queue waiting for having lunch. This situation can be defined as:

- System Φ: Line of persons waiting to be served
- Protocol ρ: (i) Count the number of persons in the queue every half an hour, and (ii) record this number indexed by the observation time.
- The attributes are natural numbers, i.e. $a[k] \in \mathbb{N}$, with indices $k \in \{10:00, 10:30, \ldots, 14:30\}$.
- The random variable for the kth observation is: $X(k) = a[k]$
- The stochastic process is defined as $\{X(10:00), \ldots, X(14:30)\}$.

Figure 3.9 exemplifies this random process considering three different realizations. Each different line refers to a different day and represents the measured

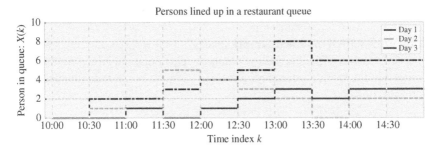

Figure 3.9 Example of a random process. The number of persons in a queue, measured every 30 minutes, from 10:00 to 14:30. Three different realizations were simulated.

number of persons in the queue following the proposed observation time sequence (the observation times are represented by the dotted vertical lines).

The different observations that form a stochastic process can be composed of iid random variables as in Example 3.5. However, this is not always the case, and thus, it is important to define random processes in terms of how a given observation is related to past observations of the same process. We will follow [3] and define the following types of stochastic processes.

Definition 3.9 *Types of stochastic processes.* Consider a process with N elements $\{X(0), X(1), \ldots X(N-1)\}$, where each random variable is associated with its own sample space $S(k)$ with $k = 0, \ldots, N-1$. We have the joint distribution: $P(X(0) = x_0, \ldots, X(N-1) = x_{N-1})$ and the conditional probability $P(X(N) = x_N \mid x_0, x_1, \ldots, x_{N-1})$. We can now define different types of stochastic processes as follows.

- **Memoryless:** $P(X(N) = x_N) = P(X(N) = x_N \mid x_0, x_1, \ldots, x_{N-1})$. In plain words, the present outcome does not depend on past realizations of the process.
- **Markov processes:** $P(X(N) = x_N \mid x_{N-1}) = P(X(N) = x_N \mid x_0, x_1, \ldots, x_{N-1})$. In other words, the present outcome only depends on the last outcome; all the previous outcomes are irrelevant.
- **Path- or history-dependent processes:** $P(X(N) = x_N \mid x_0, x_1, \ldots, x_{N-1}) = f(x_0, x_1, \ldots, x_{N-1}, x_N)$, i.e. the present outcome depends on the history, and thus, the $P(X(N) = x_N)$ is a generic function f of all prior outcomes. This is the more general class of processes.

There is also another interesting classification of processes that are related to variations in the sample spaces. They are presented next.

- **Reinforcement processes:** $S(N) = f(x_0, x_1, \ldots, x_{N-1})$. In this case, the sample space $S(N)$ from where the random variable $X(N)$ will be selected is a generic function f of the previous outcomes by positively or negatively reinforcing outcomes that were successful before.

- **Processes with dynamical sample spaces:** $|S(N)| = f(x_0, x_1, \ldots, x_{N-1})$ where $|\cdot|$ denotes the cardinality (i.e. the number of elements) of a given set. In those processes, the sample space may increase or decrease as a generic function f of previous outcomes.

Example 3.13 *Examples of different types of stochastic processes.* Consider different observation protocols or experiments related to a random drawing of tiles from a dominoes set.

- **Memoryless:** A protocol defined in Example 3.5, where the selected tile is returned to the bank to be used for the next draw. Each outcome is independent of the other outcomes.
- **Markov process:** The same protocol is used but a new experiment is defined to count the difference between the number of times that 5 and 6 are observed. The probability that the random variable will assume in the next observation for a given value only depends on its current value, and how the new observation will change state. Another similar case, but simpler, is the difference between how many times 5 and 6 are observed as outcomes of sequential dice rolls, exemplified by a Markov chain presented in Figure 3.10.
- **Path- or history-dependent process:** Consider a real dominoes game being played by two persons. If we define as a random variable the last tile put in the table, the stochastic process can be defined as the sequence of tiles being used by the players. This random process depends on the whole history of the game actions.
- **Reinforcement process:** Consider a different protocol where the tile that is selected is positively reinforced in the following way: (i) a tile is selected and removed, (ii) another random tile is randomly removed from the dominoes set, and (iii) two identical tiles from the one that was first selected are returned. In this case, the probability that the same tile will be observed in the next stage increases, positively reinforcing the chances of its occurrence.

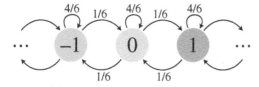

Figure 3.10 Markov chain representing a random variable X defined by the difference between the numbers of occurrences of 5 and 6 in sequential dice rolls. At each roll, the difference can grow by one if the outcome is 5 with probability 1/6, decrease by one if the outcome is 6 with probability, or be in the same state with probability 4/6.

- **Processes with dynamical sample spaces:** This case can be exemplified by applying the same observation process used in the memoryless case, but now without returning the selected tile. Therefore, the sample space dynamically decreases during the stochastic process.

This ends the proposed brief review of probability theory. Now that we are better equipped to mathematically characterize uncertainty in processes and systems, we can move on to the next chapter, where we will define *information* as a concept that indicates *uncertainty resolution*.

3.5 Summary

This chapter introduced the mathematical theory of probability as a formal way to approach uncertainty in systems or processes. By using intuitive examples, we navigated through different concepts like observation process, random variables, sample space, and probability function, among others. The idea was to set the basis for understanding the cyber domain of cyber-physical systems, which is built upon *information* – a concept closely associated with uncertainty. As a final note, we would like to stress that the present chapter is an extremely brief introduction, and thus, interested readers are strongly suggested to textbooks in the field, such as [2]. Another interesting book is [3], where the authors build a theory for complex systems using many fundamental concepts of probability theory. In particular, its second chapter provides a pedagogical way to characterize random variables and processes, including well-known probability distributions and stochastic processes. Finally, reading the original work by Kolmogorov [1] may be also a beneficial exercise.

Exercises

3.1 Players drawing two times the same five tiles. Example 3.3 analyzed the situation where one player is the first one to get five tiles out of the full set containing 28 tiles. The task here is to evaluate other situations.

(a) Compute the chances of the player to get the same five tiles considering that he/she draws after the first one has drawn his/her five tiles.

(b) Consider that the drawing is performed in a sequential manner so that one player gets one tile first, then the other player gets the second tile, and so on. Compute the chances of the first player of the sequence of getting the same five tiles regardless of the other player's hand.

(c) Compute the chances of the second player of the sequence of getting the same five tiles regardless of the other player's hand.

(d) Consider the situation where both players get the same hand. Does the drawing protocol affect the chances of the outcome?

3.2 Formal analysis to compute probability. The task in this exercise is to practice the formalization proposed in Definitions 3.2 and 3.3 and illustrated in Example 3.5.

(a) Formalize the results obtained in Example 3.2, computing the probability associated with the events described there.

(b) Do the same for Example 3.3.

3.3 Maxwell's demon operation. Consider Example 3.10. The task is to formalize the Maxwell's demon experiment based on probability theory and illustrate the equilibrium macrostates before and after the demon's intervention.

(a) Formalize the experiment following Definitions 3.2 and 3.3.

(b) Plot the Maxwell–Boltzmann distribution introduced in (3.7) for $a = 1$, which is the equilibrium state before the demon's intervention.

(c) Find the separation threshold value x_{th} that leads to $F_X(x_{th}) = 0.5$.

(d) Plot an estimation of the Maxwell–Boltzmann distribution for the two new equilibrium states. [*Hint*: do not compute the new temperature, just think about the effect of the demon's intervention to guess the new values of the constant a.]

3.4 Markov chain. The task is to extend the Markov process presented in Figure 3.10 to the dominoes case presented in Example 3.5, considering a random variable X that counts the difference between the number of times that 5 and 6 are observed (note that if [5,5] or [6,6] are observed, the state will increase by two or decrease by two, respectively, while [5,6] does maintain the process in the same state).

(a) Formalize the experiment following Definitions 3.2, 3.3, and 3.6.

(b) Prove that the difference between the number of times that two is a Markov process, i.e. $P(X(N) = x_N) = P(X(N) = x_N \mid x_{N-1})$.

(c) Find the transition probabilities $P(X(N) = x_N \mid x_{N-1})$.

(d) Represent this stochastic process as a Markov chain.

References

1 Kolmogorov A. Foundations of the Theory of Probability: Second English Edition. Courier Dover Publications; 2018.

2 Leon-Garcia A. Probability, Statistics, and Random Processes For Electrical Engineering. Pearson Prentice Hall; 2008.

3 Thurner S, Hanel R, Klimek P. Introduction to the Theory of Complex Systems. Oxford University Press; 2018.

4 Kroese DP, Brereton T, Taimre T, Botev ZI. Why the Monte Carlo method is so important today. Wiley Interdisciplinary Reviews: Computational Statistics. 2014;6(6):386–392.

4

Information

In the previous chapter, the foundations of probability theory were presented as a way to characterize uncertainty in systems or processes. Despite this fact, we have not provided any direct quantification of uncertainty in relation to the random variables or processes defined by the different probability distributions. This chapter follows Adami's interpretation of Shannon's seminal work [1], summarized in a friendly manner in [2], where the *entropy* function measures the uncertainty in relation to a given observation protocol and the experiment. Therefrom, the informative value of events – or simply *information* – can be mathematically defined. After [1], this particular branch of probability theory applied to state the fundamental limits of engineered communications systems emerged as an autonomous research field called Information Theory.

Our objective here is not to review such a theory but rather to provide its very basic concepts that have opened up new research paths from philosophy to biology, from technology to physics [3]. To this end, we will first establish what can be called information in our theoretical construction for cyber-physical systems, eliminating possible misuses or misinterpretations of the term, and then present a useful typology for information. This chapter is an attempt to move beyond Shannon by explicitly incorporating *semantics* [3, 4] but still in a generalized way as indicated by [2, 5–7].

4.1 Introduction

Information is a term that may indicate different things in different contexts. As discussed in previous chapters, this might be problematic for scientific theories, where concepts cannot be ambiguous. To clean the path for defining information as our theoretical concept, we will first critically analyze its dictionary definitions, which are presented next.

Cyber-physical Systems: Theory, Methodology, and Applications, First Edition. Pedro H. J. Nardelli.
© 2022 The Institute of Electrical and Electronics Engineers, Inc. Published 2022 by John Wiley & Sons, Inc.

Definition 4.1 *Information in [8].*

1a (i): knowledge obtained from investigation, study, or instruction, (ii): intelligence, news, (iii): facts, data

1b: the attribute inherent in and communicated by one of two or more alternative sequences or arrangements of something (such as nucleotides in DNA or binary digits in a computer program) that produce specific effects

1c (i): a signal or character (as in a communication system or computer) representing data (ii): something (such as a message, experimental data, or a picture) which justifies change in a construct (such as a plan or theory) that represents physical or mental experience or another construct

1d: a quantitative measure of the content of information specifically: a numerical quantity that measures the uncertainty in the outcome of an experiment to be performed

2 : the communication or reception of knowledge or intelligence

3 : the act of informing against a person

4 : a formal accusation of a crime made by a prosecuting officer as distinguished from an indictment presented by a grand jury

These different definitions indicate that the term "information" can be used in different domains, from very specific technical ones like 1b, 1c, 1d, or 4 to more general ones like 1a, 2, and 3. Information is then related to some kind of new knowledge or to some specific kind of *data* in technical fields. In some cases, information can be seen as something that brings novelty in knowledge, whereas in other cases, information is a sort of representation of data or uncertainty. Although some definitions might be more appealing for our purpose, we would like to propose a different one that will allow us to formally establish a concept for information.

Definition 4.2 *Information.* Information refers to a trustworthy uncertainty resolution.

With this definition, we are ready to start. Our first task is to differentiate what is data and what is information, as well as their interrelation.

4.2 Data and Information

Data and information are usually considered synonymous, as indicated in Definition 4.1. However, if we follow Definition 4.2, we can deduce that not all data are informative because some data may not be relevant for resolving the uncertainty related to specific processes. For example, the outcome of the lottery is not informative with respect to the weather forecast, but it is informative for the tax office. Information is then composed of data, but not all data are information. To formalize this, we will state the following proposition.

Proposition 4.1 *Data and information.* Considering a particular system or process Φ, let D denote an infinite countable set of all possible (known and unknown) data d_i, with $i = 0,1,\ldots$, that are used to characterize Φ. The set D contains, for instance, statements, measurements, images, qualities, attributes, and opinions that can be (i) structured or not, (ii) meaningful or not, and (iii) trustworthy representations or not. Let I denote a subset of D that potentially contains *information* about Φ composed by d_is that are (i) syntactically structured, (ii) semantically meaningful, and (iii) potentially trustworthy representations of Φ. Therefore, $I \subset D$ implies that not all data about Φ have the potential to be information.

In other words, this proposition indicates that, to potentially be information, data must be interpretable (i.e. follow a syntax and be meaningful). Such meaningful structured data also need to carry a *true* (correct, trustworthy) reflection of Φ. If such data decrease the uncertainty about some well-defined aspect of Φ, then the *data are information*. Information is then a concept that depends on the uncertainty *before* and *after* it is acquired.

The following two examples illustrate these ideas.

Example 4.1 *Lottery.* Once a month, there is a lottery with very simple rules: (i) there are 61 balls in an urn numbered from 0 to 60, (ii) six balls are randomly selected one by one, and (iii) whoever has the six numbers wins the lottery. This is a chance game that can be analyzed following the concepts presented in Chapter 3. In this game, the relevant data are the numbers written on the balls, which have syntax and semantics with respect to the game rules. For a person with a lottery ticket who is following the draw in real time on TV, the numbers that appear on the TV screen are a correct representation of the number written on the ball. Each ball that is drawn decreases the uncertainty about the winning numbers. Therefore, each number that appears on TV satisfies the conditions to be considered information.

Example 4.2 *Lecture.* Consider a situation that a given teacher is lecturing about the relation between data and information. To understand the lecture, the students and the teacher should share a set of norms (e.g. English syntax) in order for the sounds and symbols to be mutually understandable (i.e. having meaning). Data such as sounds, symbols, and images are generated by the teacher to demonstrate a relation between data and information. Such data must be articulated to form a consistent theoretical construction so that the lecture is a trustworthy representation of the theory being presented. The lecture is the means to transfer the data from the teacher to the students. After the lecture, if a given student learned a new knowledge, then the data generated by the teacher can be called information with respect to that student.

As those examples indicate, data and information may appear in different forms, and thus, it is worth using a typology to support their analysis [3]. Data may be either *analog* or *digital*. For example, the voice of a person in the room is analog, while the same voice stored as a computer file in an mp3 format is digital. Analog data may be recorded in, for example, vinyl discs. Computer files associated with audio are not recorded in this way; they are actually encoded by state machines via logical bits, i.e. 0s and 1s that can be physically represented by circuits [9]. Bits can then be represented semantically by true or false answers [10], mathematically based on Boolean algebra [11], and physically by circuits. It is then possible to construct machines that can [p. 29][3] (…) *recognize bits physically, behave logically on the basis of such recognition and therefore manipulate data in ways which we find meaningful.* As we are going to see later, this fact is the basis of the *cyber* domain of *every cyber-physical system.*

Beyond this more fundamental classification, we can list the following types of data (adapted from [4]):

- **Primary data:** Data that are directly accessible as numbers in spreadsheets, traffic lights, or a binary string; these are the data we usually talk about.
- **Secondary data:** The absence of data that may be informative (e.g. if a person cannot hear any sound from the office beside his/hers, this silence – lack of data – indicates that no one is there).
- **Metadata:** Data about other data as, for instance, the location where a photo was taken.
- **Operational data:** Data related to the operations of a system or process, including its performance (e.g. blinking yellow traffic lights indicate that the signaling system that supports the traffic in that street is not working properly).
- **Derivative data:** Data that are obtained through other data as indirect sources; for example, the electricity consumption of a household (primary data) also indicates the activity of persons living there (derivative data).

It is easy to see that data can be basically anything, including incorrect or uninterpretable representations and relations. This supports our proposal of differentiating data and information, as stated in Proposition 4.1. In this case, classifying data according to the proposed typology may be helpful in the process of identifying what has the potential to be considered information. Nevertheless, Examples 4.1 and 4.2 indicate how broad these concepts are. The first example can be fully described using probability theory, while the second cannot (although some may argue that it is always possible to build such a model). This is the challenge we should face.

Our approach will be to follow Proposition 4.1, and thus, establish generic relations involving data, information, and uncertainty with respect to well-defined scenarios. The following proposition states such relations.

Proposition 4.2 *Uncertainty and information.* Consider a particular system or process Φ, and its respective sets D and I, as introduced in Proposition 4.1. Let $X \in \mathcal{X}$ be a specific characteristic of Φ where $\mathcal{X} \subseteq I$ is a set that contains all valid values that X can assume, and $H(X)$ is a generic function that quantifies the uncertainty about X so that $H : \mathcal{X} \to \mathbb{R}$. If we similarly define another characteristic $Y \in \mathcal{Y}$ of Φ with $\mathcal{Y} \subseteq I$, then:

$$H(X|Y) \leq H(X), \tag{4.1}$$

$$H(Y|X) \leq H(Y), \tag{4.2}$$

where $H(X|Y)$ can be read as "the uncertainty of X given the knowledge of Y." The equality in both equations is obtained when X and Y are independent from each other (i.e. X is not informative in relation to Y, and vice versa).

The mutual information between X and Y, denoted by $I(X;Y)$, can be computed as:

$$I(X;Y) = H(X) - H(X|Y) = H(Y) - H(Y|X). \tag{4.3}$$

If $X \not\subseteq I$ or $Y \not\subseteq I$, the relations (4.1), (4.2), and (4.3) are not applicable, and thus, do not necessarily hold.

These are general statements that are usually presented in a different manner without differentiating data and information. As we will see in the next section, this may be problematic. At this point, we will illustrate this statement with the following example.

Example 4.3 *Lottery and uncertainty.* Let us consider the lottery presented in Example 4.1. The numbered balls have three colors: black for even numbers, white for odd numbers, and red for zero. Following Propositions 4.1 and 4.2, we can state the following.

- Φ is the aforementioned lottery process.
- $D = \{$ "white ball", "black ball", "red ball", "0", ..., "60", "fair", "unfair", "α", "square", "12BH24F", ...$\}$.
- $I = \{$ "white ball", "black ball", "red ball", "0", ..., "60", "fair", "unfair", ...$\}$.

- X is the number written on the ball so that $\mathcal{X} = \{\,\text{"0"}, \ldots, \text{"60"}\,\}$.
- Y is the color of the ball so that $\mathcal{Y} = \{\,\text{"white ball", "black ball", "red ball"}\,\}$

Let us consider the first draw where any of the sixty-one balls can be taken randomly. We assume that this process is fair, which is presupposed information about the lottery. In a given draw, it is possible to measure the uncertainties $H(X)$ and $H(Y)$ in relation to X and Y, which are potentially informative characteristics of the lottery process Φ. Intuitively, if X is known, there is no uncertainty about Y and then $H(Y|X) = 0$. On the other hand, if Y is known, the uncertainty of X decreases, but it is not zero. By manipulating (4.3), we have

$$I(X; Y) = H(Y) - H(Y|X) = H(Y), \text{ and}$$

$$I(X; Y) = H(X) - H(X|Y), \text{ then}$$

$$H(X|Y) = H(X) - H(Y).$$

In this case, $H(X) > H(Y)$ since $H(X|Y) > 0$.

On the other hand, if we assume that the drawing process is unfair, the equations stated above might not be valid because it imposes another uncertainty on our problem. This uncertainty is of a different level: it is not specified, thus making unfeasible any sound analysis of the uncertainty related to X and Y. We can say the information that the lottery process is unfair, without further specification, breaks the possibility of evaluating $H(X)$ and $H(Y)$. A nonspecified unfairness in the process (which is a "known known") converts a mathematically tractable process defined by "known unknowns" into an intractable one composed of "unknown unknowns."

Let us now consider the elements of the set D that are not elements of $\mathcal{I}$, denoted as $D - \mathcal{I}$. They are irrelevant to the problem because, by definition, they cannot be used to resolve any uncertainty by being uninterpretable (e.g. the string "12BH24F"), or meaningless (e.g. the symbol "α"), or cannot provide a trustworthy representation of the process Φ (e.g. the characteristic "square").

As in probability theory, the evaluation of uncertainty about a given aspect of a system or process depends on a few certainties that are considered as givens. The example above demonstrates that an undefined uncertainty about the observation process that is associated with the aspect whose uncertainty we are willing to evaluate makes such an analysis unfeasible. Even in this situation, we could take additional assumptions and evaluate the extreme scenarios. The following proposition states additional properties of the uncertainty function H.

Proposition 4.3 *Properties of the uncertainty function.* Consider the characteristic X of the process or system Φ and the function $H(X)$ that measures its uncertainty, as defined in Proposition 4.1. The lower and upper limits of $H(X)$ are, respectively: (i) $H(X) = 0$ when there is no uncertainty about X, and (ii) $H(X) = H_{max}$ where $H_{max} \geq 0$ refers to the maximum uncertainty about X obtained when all possible outcomes $X = x$ with $x \in \mathcal{X}$ are equiprobable. If the set $\mathcal{X}$ is finite, then $H_{max} < \infty$.

It is intuitive that if there is no uncertainty in the characteristic X, there is no uncertainty and $H(X) = 0$. Considering that the set $\mathcal{X}$ is well defined, then the worst-case scenario is when all its elements are equally probable to be observed (i.e. they have the same chances to be the outcome of the observation process that will determine the actual value x that the characteristic X assumes). With these properties in mind, it is possible to mathematically define $H(X)$ as follows.

Definition 4.3 *Information entropy as the uncertainty function.* Consider the following function $X(a)$ associated with observable attributes $a \in \mathcal{A}$ of the process or system Φ so that $X : \mathcal{A} \rightarrow \mathbb{R}$. Without loss of generality, we can assume that $X(a)$, or simply X, is a random variable that represents the characteristic of the process Φ as defined in Proposition 4.1. In this case, the sample space of X is $S_X \equiv \mathcal{X}$ and its probability mass function is $p_X(x) = P(X = x)$, following the definitions presented in Chapter 3. The uncertainty function $H(X)$ – called Shannon or information entropy – is [1]:

$$H(X) = - \sum_{i \in S_X} P(X = i)\log_2 P(X = i), \tag{4.4}$$

which is measured without loss of generality in bits (i.e. binary digits).

The maximum uncertainty H_{max} is obtained if all the elements of the sample space S_X are equally probable. If we assume that S_X has N elements, then $P(X = x) = 1/N$ for all $x \in S_X$. Hence,

$$H_{max} = \log_2 N, \tag{4.5}$$

which depends only on the cardinality of the sample space.

It is interesting to note that (4.4) can be derived from basic axioms as presented in [5] or in [10]. The definition of joint and conditional entropy functions can be readily defined following (4.4) by using the joint and conditional probability mass functions. Figure 4.1 presents a pedagogical illustration of how the conditional and joint entropy are related in the case of two variables X and Y as presented in Proposition 4.3. Note this is a usual visual representation of mathematical concepts that might be misleading in many ways.

It is also possible to extend this definition to continuous random variables as already presented in [1], but the definition would be more complicated since it would involve integrals instead of summations (the principles are nevertheless the

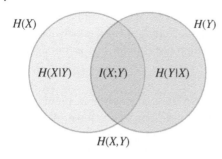

Figure 4.1 Relation between entropy functions and mutual information for two random variables X and Y.

same). Another important remark is related to the term "entropy" because it has been used in different scientific fields such as physics, information theory, statistics, and biology. As clearly presented in [5], the formulation of entropy presented in Definition 4.3 is always valid for memoryless and Markov processes, being very similar regardless of the research field; the only differences are the way it is presented and possible constants for normalization (mainly in physics). For complex systems that are related to, for instance, history-dependent or reinforcement processes, the entropy function needs to be reviewed because the axioms used to derive (4.4) are violated. These alternative formulations are presented in [5].

Although important, those advances in both information theory and entropy measures for complex systems are well beyond the scope of this book. To finish this section, we will provide a numerical example to calculate the uncertainty of the lottery previously analyzed.

Example 4.4 *Uncertainty and information in the lottery.* Consider the lottery presented in Examples 4.1 and 4.3. We can then relate the characteristics of interest to two attributes:

- X is the number written on the ball so that $\mathcal{X} = \{$"0", ..., "60"$\}$, and
- Y is the color of the ball so that $\mathcal{Y} = \{$"white ball", "black ball", "red ball"$\}$.

As an example, let us evaluate a random draw of one ball assuming a fair process so that we can analyze those two characteristics as random variables, starting with X. The probability mass function is the uniform distribution over the sixty-one elements of the sample space, and thus,

- $p_X(x) = 1/61$;
- $H(X) = H_{\max,X} = \log_2 61 = 5.93$ bits.

This number measured in bits can be seen as the average minimum number of yes-or-no questions that need to be answered to solve the uncertainty related to the outcome of X. For example, we can ask the question: *is the outcome $X = x$ between 0 and 30?* If yes, we narrow down the question using the limits 0 and 15; if no, we know that $X = x$ is in the range between 31 and 60, and we narrow down the

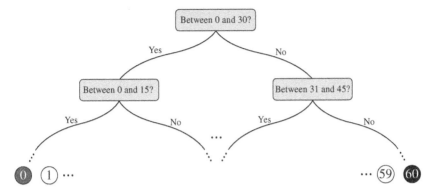

Figure 4.2 Tree representing yes-or-no questions to resolve the uncertainty in the lottery.

question using the limits 31 and 45. By decreasing the sampling space in this way, we can solve the uncertainty after 5.93 questions on average. Figure 4.2 illustrates this process.

In relation to Y, we know beforehand that black balls are for even numbers, white for odd, and red for zero. Then,

- $p_Y(y) = 30/61$ if $y \neq 0$, $p_Y(y) = 1/61$ if $y = 0$;
- $H(Y) = -30/61\log_2(30/61) - 30/61\log_2(30/61) - 1/61\log_2(1/61) = 1.1$ bits;
- $H_{\max,Y} = \log_2 3 = 1.58$ bits (this is the case when no knowledge is assumed beforehand, i.e. the worst-case scenario for sample space with three elements);
- $H_{\max,Y} - H(Y) = 0.48$, which indicates the *information* contained in the prior knowledge about Y.

We also know that if the value of X is known, there is no uncertainty in the variable Y. Then:

- $H(Y|X) = 0$ bits;
- $I(X;Y) = H(Y) - H(Y|X) = H(Y) = 1.1$ bits;
- $H(X|Y) = H(X) - H(Y) = 4.83$ bits.

We can interpret this as follows: the first two equations state that all information contained in Y is contained in X, and that the knowledge of Y decreases, on average, by 1.1 the number of yes-or-no questions needed to be asked to resolve this uncertainty.

4.3 Information and Its Different Forms

In the previous section, data and information as concepts were distinguished, the second being a subset of the first whose elements are interpretable and can resolve the uncertainty related to the specific characteristics of a given system or process. We have also introduced a way to mathematically formalize and quantify uncertainty and information by entropy, which is valid for some particular cases. This section will provide another angle on the analysis of information that focuses on the different forms in which that information exists. Following [3], information can be: (a) mathematical, (b) semantic, (c) biological, or (d) physical.

4.3.1 Mathematical Information and Communication

Shannon, in his seminal paper [1], established *a mathematical theory of communications*, where

> (...) the fundamental problem of communication is that of reproducing at one point either exactly or approximately a message selected at another point. Frequently the messages have meaning; that is they refer to or are correlated according to some system with certain physical or conceptual entities. These semantic aspects of communication are irrelevant to the engineering problem. The significant aspect is that the actual message is one selected from a set of possible messages. The system must be designed to operate for each possible selection, not just the one which will actually be chosen since this is unknown at the time of design.

The engineering problem is depicted in Figure 4.3.

Shannon mathematically described the data messages regardless of their semantics and gave them a probabilistic characterization by random variables X and Y that are associated with the transmitted and received messages, respectively. To do so, Shannon introduced the entropy function to measure the uncertainty of the random variables and then maximize the mutual information between them; these

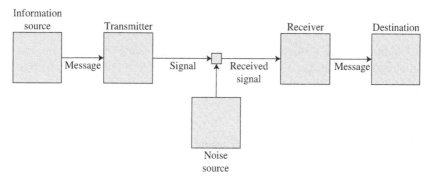

Figure 4.3 Diagram of a general communication system by Shannon.

equations were introduced in the previous section. This paper by Shannon is considered the starting point of information theory, a remarkable scientific milestone. The main strength of Shannon's theory is the generality of the results, establishing the fundamental limits of communication channels and also data compression (e.g. [10, 12]). We can say that this generality is also its main weakness because it eliminates the semantics of particular problems to be solved. In simple words, Shannon states the limits of an error-free transmission of information that no past, existing, or future technology can surpass, which might be of less importance at the semantic level where many problems related to the information transmission can be suppressed (for example, a missing data point related to the temperature sensor can be fairly well estimated from the past) while new ones appear. Semantic problems are usually history-dependent processes also associated with changes in the sample spaces so that Shannon's key results need to be revisited [5, 7]. More details about semantic information are given next.

4.3.2 Semantic Information

Semantics refers to the meaning of data and therefore, of information. Semantic information can be classified as [3]:

- **Instructional:** Data or information that prescribes actions in a "how to" manner. For example, a recipe that states the procedure to make a chocolate cake.
- **Factual:** Data that can be declared true or false. Information is then defined as data that are true while misinformation is false data that are unintentional and disinformation false data that are intentional. For example, a scientific paper with new results can be considered factual information. However, it may also be factual misinformation when the measurement devices have not been calibrated and the authors are not aware of it, and thus, the results are unintentionally false. Making fake results to publish a paper is a case of factual disinformation.
- **Environmental:** Data or information that is independent of an intelligent producer. For example, a tree outside an office indicates the season of the year (green leaves, no leaf, fruits, flowers, etc.). Engineering systems are also designed to be a source of environmental information, such as a thermometer. In general, environmental information is defined as the coupling of two (or more) processes where observable states of one or more processes carry information about the state of other process(es).

In summary, semantic data must be judged with respect to its associated meaning or its purpose, and thus, it is different from – and in many senses more general than – Shannon's mathematical theory. For instance, an error-free communication in Shannon's sense might be associated with misinformation, like in the case when a perfect communication link transmits the data of a poorly calibrated sensor that misinforms the application that will use such data. As Shannon himself has indicated in his paper, processes like "natural language" could be analyzed by his probabilistic theory but without any ambition to fully characterize it. Natural language can be seen as an evolutionary (history-dependent) process that is built upon semantics that cannot be reduced to syntactical/logical relations alone.

While semantics is associated with meaning and hence with the social domain, we could also analyze how life is associated with data and information.

4.3.3 Biological Information

The growing scientific understanding of different biological processes including physiology, pathology, genetics, and evolution of species indicate the necessity of "data storage, transmission, and processing" for the production and reproduction of life in its different forms [13]. Biological information broadly refers to how the organization in living beings could (i) emerge from chemical reactions, (ii) be maintained allowing reproduction of life, and (iii) be associated with the emergence of new species [14]. Different from semantic information associated with socially constructed meanings, biological information concerns data that are transmitted by chemical signals that are constitutive of living organisms that evolve over time.

As indicated in [13], Shannon's Information Theory has had a great influence on biology. Even more interesting is the fact that Shannon himself, before moving to Bell Labs, wrote his Ph.D. thesis about *theoretical genetics* [15]. Biological information can be used as a tool to understand the emergence of new species, cell differentiation, embryonic development, the effects of hormones in physiology, and nervous systems, among many others. It is worth noting that, despite the elegance and usefulness of some mathematical models, the peculiarity of biological objects and concepts must be preserved. In this sense, biological information is still a controversial topic [13]. Nevertheless, we would like to stress the importance of the different existing approaches that use the concept of biological information to develop technologies in, for instance, biochemistry, pharmacology, biomedical engineering, genetic engineering, brain–machine or human–machine interfaces, and molecular communications.

We follow here [3] by arguing that biological information deserves to be considered in its own particularities. Biological information cannot be reduced or derived from Shannon's mathematical information, or semantic information. On the contrary, the scientific knowledge of biological information needs to be constructed with respect to living organisms, even though mathematical and semantic information may be useful in some cases.

4.3.4   Physical Information

As indicated by the Maxwell's demon examples in the previous chapters, there seems to be a fundamental relation between physical and computational processes. Storing, processing, transmitting, and receiving data are physical processes that consume energy, and thus, their fundamental limits are stated by thermodynamics. Landauer theoretically proved in [16] and Berut et al. experimentally demonstrated in [17] that the erasure of a bit is an irreversible process, and thus, dissipates heat. Similarly, Koski et al. proved in [18] that the transfer of information associated with their experimental setting of Maxwell's demon dissipates heat. The dissipated heat and the information entropy are then fundamentally linked: Landauer's principle states the lower bound of irreversible data processing. In Landauer's interpretation [19], this result indicates that "information is physical."

The fundamental relation between energy and information motivates the study of reversible computation, and also ways to improve the energy efficiency of existing irreversible computing methods by stating their ultimate lower bound. Despite its evident importance, such a fundamental result has had, until now, far less impact on technological development than Shannon's work. However, this may change because of large-scale data centers with their extremely high energy demand not only to supply the energy to run the data processing tasks but also for cooling pieces of hardware that indeed dissipate heat. What is important to understand here is that any data process must have a material implementation, and thus, requires energy.

In this sense, the limits stated by physical information are valid for all classes of information. Both biochemical processes related to biological information and interpretative processes related to semantic information consume energy, and hence, have their fundamental limits established by Shannon and Landauer. Note that the fundamental limits are valid even for processes whose mathematical characterization defined by the mathematical information is not possible.

The concept of physical information also brings another interesting aspect already indicated in the previous sections: information is a relational concept defined as a trustworthy uncertainty resolution, which is always defined with respect to a specific characteristic of a process or a system. It is then intuitive to think that information should refer, directly or indirectly, to something that materially exists. This is the topic of our next section.

4.4   Physical and Symbolic Realities

The scientific results related to information, whose foundations were just presented, are many times counterintuitive and against the common sense of a given

historical period. As discussed in Chapter 1, this may lead to a philosophical interpretation (and distortion) of scientific findings. For example, there is a good deal
of heated debates about the nature of information (i.e. metaphysics of information). For example, the already cited works [2–4, 16, 20] have their own positions
about it.

In this book, we follow a materialist position, and thus, we are not interested
in the nature of information; for us, *information exists*. In our view, this is the
approach taken by Shannon, and only in this way science can evolve. Despite our
critical position about Wiener's cybernetics presented in Chapter 1 [21], he precisely stated there how information should be understood:

> The mechanical brain does not secrete thought "as the liver does bile," as
> the earlier materialists claimed, nor does it put it out in the form of energy,
> as the muscle puts out its activity. Information is information, not matter
> or energy. No materialism which does not admit this can survive at the
> present day.

Information exists, as matter and energy do, but it operates in its own reality, which
we call here *symbolic reality*. The relation between symbolic and physical realities
is given by the following proposition.

Proposition 4.4   *Physical and symbolic realities.* Physical (material) processes exist and constitute the physical reality. Data processes can only exist supported, directly or indirectly, by the physical reality. Therefore, data processes are
necessarily physical, but they originate new relations, referred to as *symbolic relations*, which constitute another domain denominated as *symbolic reality*. In specific cases dealing with engineered computing apparatuses, the symbolic relations
are called *logical relations*, and the symbolic reality is called *cyber reality*.

In other words, this proposition tells that data processes are physical and constitute relations that are not only physical but also symbolic. Information as a
product of data processes is then physical and exists in the symbolic reality. It is
also possible to define processes that are constituted in the symbolic reality related
to symbolic relations. These processes can be classified into levels with respect to
the physical reality, as stated in the following definition.

Definition 4.4   *Level of processes.* Level 0 refers to the physical reality. Level
1 refers to symbolic processes that involve data directly obtained from the physical
reality associated with level 0 processes. Level 2 refers to symbolic processes that
involve data obtained from the symbolic reality whose process of highest level is 1.

Level *N* refers to symbolic processes that involve data obtained from the symbolic reality whose process of highest level is *N* − 1.

Example 4.5 *Car in a highway.* A car moving on a highway can be analyzed as a level 0 process. The car speedometer can be seen as part of a level 1 process because it measures the speed that the car moves on the highway: the speed is the data acquired from a level 0 process and employed in a level 1 process. The speedometer also has a transmitter device that sends via a wireless channel a message to the traffic authority whenever the car exceeds the speed limit of the highway. The traffic authority charges a fine (which can be seen as a level 2 process) if it receives three messages (data acquired from the level 1 process) from the same car in a period of twelve months. The government uses the number of fines (seen here as a level 2 process) to study the statistics of the traffic in a given region (which can be analyzed as a level 3 process). This is depicted in Figure 4.4

From this very simple example, it is possible to see that the space of analysis could grow larger and larger because symbolic processes are unbounded. Note that higher levels tend to require more data processing, and thus, consume more energy. When the physical reality is mapped into the data forming a symbolic reality, the number of ways that these acquired data can be combined grows to infinity. However, as stated in Proposition 4.1, data that have the potential to become information constitute a very restricted subset of all data about a specific process or system. Given these facts, there are limits in the symbolic relation that cannot be passed, as presented in the following proposition.

Proposition 4.5 *Energy and information limits of the symbolic reality.* Symbolic reality always involves processes (e.g. data acquisition, storage, and

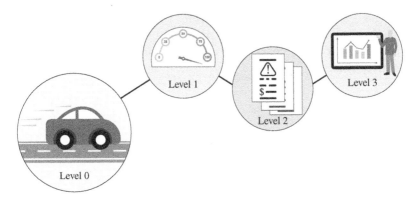

Figure 4.4 Car in a highway and level of processes.

manipulation) that require energy, which is limited. Information presupposes trustworthy structured data of a given reality, regardless of the level of the process. Therefore, the number of informative combinations of data is limited.

In this case, all data processing is limited by energy and, more restrictively, the requirements that the data must fulfill to become information constrain even more what could potentially be an informative symbolic reality, which is the foundation of the cyber domain of cyber-physical systems. In the cyber domain, the symbolic relation is usually called logical relations in contrast to physical relations. The next step is to understand how physical and symbolic relations can be defined through *networks*.

4.5 Summary

This chapter introduced the concept of information. We defined it as a subset of all data about a specific process or system that is structured and meaningful, and thus, capable of providing a trustworthy representation. Information is then related to uncertainty resolution, which always involves the processing of data. For some cases, we explained how to measure the uncertainty from a probabilistic characterization of random variables (as presented in Chapter 3) by Shannon's entropy. Therefrom, we presented how to compute the mutual information between two random variables, as well as their conditional and joint entropies. We also presented a discussion about different types of data and information, indicating the fundamental limits of processes involving manipulation of data.

For readers interested in the topic, the book by Floridi [3] provides a brief overview of the subject. The entries of the *Stanford Encyclopedia of Philosophy* [4, 13, 20] go deeper into different views and controversies related to this theme. Adami provides a speaking-style introduction of the mathematical basis of information theory [2]. However, the must-read is the paper by Shannon himself [1] where the fundamental limits of information are pedagogically stated. Landauer's paper [19] is also very instructive to think about the relation between information and thermodynamics.

Exercises

4.1 **Data and information.** The goal of this task is to explore how data and information are related by following Proposition 4.1.
   (a) Define in your own words how data and information are related.
   (b) Propose one example similar to Example 4.1.
   (c) Extend your example by proposing a list of data that are not informative.

4.2 **Entropy and mutual information.** The discussions about climate change always point to the dangers of global warming, which is usually measured by an increase of X degrees Celsius. This value X is taken as an average over different places during a specific period of time. Negationists, in their turn, argue that global warming is not true because a given city A reached a temperature Y that was the coldest in decades. In this exercise, we consider that X and Y are obtained through data processes P_X and P_Y and can be analyzed as random variables.

(a) Which data process, P_X or P_Y, has a greater level in the sense presented in Definition 4.4? Why?

(b) Consider a random variable Z associated with the answer of the question: *is the global warming true?* What is the relation between the entropies $H(Z)$, $H(Z|X)$, and $H(Z|Y)$?

(c) Now, consider another random variable W associated with the answer of the question: *how was the weather of city A during last winter?* What is the relation between the entropies $H(W)$, $H(W|X)$, and $H(W|Y)$?

(d) What could be said about the mutual information $I(Z;X)$, $I(Z;Y)$, $I(W;X)$, $I(W;X)$, and $I(X;Y)$?

(e) What is the problem of the argument used by negationists?

4.3 **Information and energy in an AND gate.** Consider a logical gate AND with X_1 and X_2 as inputs, and Y as output. In this case,

X_1	X_2	Y
0	0	0
0	1	0
1	0	0
1	1	1

Assume that $X_1 = 1$ is associated with a probability p_1 and $X_2 = 1$ with a probability p_2.

(a) Formalize the problem as introduced in Chapter 3 and then find the probability mass functions of X_1, X_2, and Y, as well as all conditional and joint probabilities between X_1, X_2, and Y.

(b) What are $H(X_1)$, $H(X_2)$, and $H(Y)$ assuming that $p_1 = p_2 = p$?

(c) What is the mutual information $I(X_1, Y)$ and $I(X_2, Y)$ when $p_1 = p_2 = p$?

(d) Does the AND gate operation generate heat? Why?

References

1 Shannon CE. A mathematical theory of communication. The Bell System Technical Journal. 1948;27(3):379–423.

2 Adami C. What is information? Philosophical Transactions of The Royal Society A. 2016;374(2063):20150230.

3 Floridi L. Information: A Very Short Introduction. OUP Oxford; 2010.

4 Floridi L, Zalta EN, editor. Semantic Conceptions of Information. Metaphysics Research Lab, Stanford University; 2019. Last accessed 8 January 2021. https://plato.stanford.edu/archives/win2019/entries/information-semantic/.

5 Thurner S, Hanel R, Klimek P. Introduction to the Theory of Complex Systems. Oxford University Press; 2018.

6 Li M, Vitányi P, et al. An Introduction to Kolmogorov Complexity and its Applications. vol. 3. Springer; 2008.

7 Kountouris M, Pappas N. Semantics-empowered communication for networked intelligent systems. IEEE Communications Magazine, 2021;59(6):96–102, doi: 10.1109/MCOM.001.2000604.

8 Merriam-Webster Dictionary. Information; 2021. Last accessed 11 January 2021. https://www.merriam-webster.com/dictionary/information.

9 Shannon CE. A symbolic analysis of relay and switching circuits. Electrical Engineering. 1938;57(12):713–723. Available at: http://hdl.handle.net/1721.1/11173. Last accessed 11 November 2020.

10 Ash RB. Information Theory. Dover books on advanced mathematics. Dover Publications; 1990.

11 Boole G. The Mathematical Analysis of Logic. Philosophical Library; 1847.

12 Popovski P. Wireless Connectivity: An Intuitive and Fundamental Guide. John Wiley & Sons; 2020.

13 Godfrey-Smith P, Sterelny K, Zalta EN, editor. Biological Information. Metaphysics Research Lab, Stanford University; 2016. https://plato.stanford.edu/archives/sum2016/entries/information-biological/.

14 Prigogine I, Stengers I. Order Out of Chaos: Man's New Dialogue with Nature. Verso Books; 2018.

15 Shannon CE. An algebra for theoretical genetics; 1940. Ph.D. thesis. Massachusetts Institute of Technology.

16 Landauer R. Irreversibility and heat generation in the computing process. IBM Journal of Research and Development. 1961;5(3):183–191.

17 Bérut A, Arakelyan A, Petrosyan A, Ciliberto S, Dillenschneider R, Lutz E. Experimental verification of Landauer's principle linking information and thermodynamics. Nature. 2012;483(7388):187–189.

18 Koski JV, Kutvonen A, Khaymovich IM, Ala-Nissila T, Pekola JP. On-chip Maxwell's demon as an information-powered refrigerator. Physical Review Letters. 2015;115(26):260602.

19 Landauer R, et al. Information is physical. Physics Today. 1991;44(5):23–29.

20 Adriaans P, Zalta EN, editor. Information. Metaphysics Research Lab, Stanford University; 2020. Last accessed 8 January 2021. https://plato.stanford.edu/archives/fall2020/entries/information/.

21 Wiener N. Cybernetics or Control and Communication in the Animal and the Machine. MIT press; 2019.

5

Network

So far, we have introduced the key concepts used to demarcate systems, to characterize uncertainty by probability theory, and to define data and information as the basis of cyber realities. In all the previous chapters, we have indicated that different elements under analysis are related to each other, but without explicitly defining how. This chapter introduces the basics of graph theory and network sciences, which are the disciplines providing the fundamental concepts to evaluate structured relations between different elements [1]. We will first introduce the mathematical theory of graphs, which formalizes the relation between vertices by edges. Then, we will present how graph theory is employed in other domains to evaluate particular *networks* where *nodes* are connected with each other by *links*. Different network topologies will be presented together with illustrative applications, for example, in transportation and epidemiology. We will also indicate the main limitations and problems of network sciences, which concern the relation between theoretical models and actual physical processes. This chapter provides only a brief overview of the field; for interested readers, the interactive book *Network Science* [2] is highly recommended.

5.1   Introduction

Graph theory is a field in mathematics that deals with structures using pairwise relations. The field started with the study about the *Seven Bridges of Königsberg* by Euler [3]. The problem is described as follows [4]:

> The Seven Bridges of Königsberg is a historically notable problem in mathematics. Its negative resolution by Leonhard Euler in 1736 laid the foundations of graph theory and prefigured the idea of topology. The city of Königsberg in Prussia (now Kaliningrad, Russia) was set on both sides of the Pregel River, and included two large island – Kneiphof and Lomse – which

Cyber-physical Systems: Theory, Methodology, and Applications, First Edition. Pedro H. J. Nardelli.

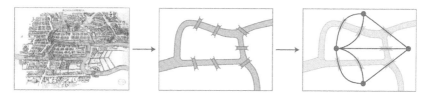

Figure 5.1 Euler's approach to prove that the problem has no solution.

were connected to each other, or to the two mainland portions of the city, by seven bridges. The problem was to devise a walk through the city that would cross each of those bridges once and only once. By way of specifying the logical task unambiguously, solutions involving either reaching an island or mainland bank other than via one of the bridges, or accessing any bridge without crossing to its other end are explicitly unacceptable. Euler proved that the problem has no solution. The difficulty he faced was the development of a suitable technique of analysis, and of subsequent tests that established this assertion with mathematical rigor.

Euler solved the problem by abstracting bridges and islands as *edges* and *vertices*, respectively. Figure 5.1 illustrates the problem. Euler's proof of impossibility of having a solution to the problem is considered the first result of graph theory, and network sciences [2, 5].

Graphs, as mathematical objects, are constituted by vertices that are connected by edges. For instance, the diagram on the right side in Figure 5.1 depicts the seven bridges problem as a graph composed of four vertices (circles) and linked by seven edges (black lines). This uniquely defines a specific topological structure of the graph. Note that, as a mathematical abstraction, the same graph may represent different real-world *networks*. The mathematical formalism of graph theory, which is also applied by network sciences, is presented next.

Definition 5.1 *Graph and network* A graph $G = (\mathcal{V}, \mathcal{E})$ is a mathematical object composed of a set of N *vertices* $\mathcal{V} = \{v_1, \ldots, v_N\}$ that are connected by L *edges*, defined by the set $\mathcal{E} = \{e_1, \ldots, e_L\}$ where $e_i = (v_j, v_h)$ so that $v_j, v_h \in \mathcal{V}$. In *undirected* graphs, the edge e_i is bidirectional, and thus, it incorporates both edges from v_j to v_h and from v_h to v_j, i.e. $(v_j, v_h) \equiv (v_h, v_j)$. In *directed* graphs, the edges are unidirectional, and thus, $(v_j, v_h) \not\equiv (v_h, v_j)$. If G refers to a real-world structure, graph, vertex, and edge are usually called *network*, *node*, and *link*, respectively.

In this book, we are not interested in studying graphs as pure mathematical objects but rather as networks constituted of either physical or logical relations, which will be used to understand cyber-physical systems in the following chapters.

Figure 5.2 Example of an undirected network with $N = 6$ nodes and $L = 7$ links.

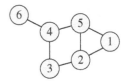

Electric power grids and computer networks are examples of networks constituted by physical relations. Markets or social media are examples of networks constituted by logical relations.

Figure 5.2 illustrates an example of an undirected network with six nodes and seven links, i.e. $N = 6$ and $L = 7$. This graph could be a representation of connections of different university buildings, or of terminals in an airport, or researchers collaborating in papers. The first two are physical networks indicating the connections of separated buildings/terminals: building/terminal 6 is connected to building/terminal 4, which is connected to buildings/terminals 3 and 5 in addition to 6, and so on. In the last case, we have a logical network of co-authors so that researcher 6 has a paper in collaboration with 4, who has also collaborated with 3 and 5, and so on.

Mathematically, this network $G = (\mathcal{V}, \mathcal{E})$ is defined by the sets:

- $\mathcal{V} = \{v_1, v_2, v_3, v_4, v_5, v_6\}$,
- $\mathcal{E} = \{(v_1, v_2), (v_1, v_5), (v_2, v_3)(v_2, v_5), (v_3, v_4), (v_4, v_5), (v_4, v_6)\}$.

The cardinality (the number of elements) of the set $\mathcal{V}$ is $N = 6$, and this number defines the *size of the network* represented by graph G. The cardinality of $\mathcal{E}$ is $L = 7$, which indicates the number of links existing in the network. Note that networks with undirected links have, in fact, $2L$ directed links; in specific cases, this must be taken into account. Sometimes, it is also possible to have edges to itself, i.e. $e_k = (v_i, v_i)$.

Another way to mathematically represent a graph is the *adjacency matrix A* with elements $a_{i,j}$, with $i, j = 1, \dots, N$ so that $a_{i,j} = 1$ or $a_{i,j} = 0$ indicating the presence or absence of a (directed) edge from vertex v_i to vertex v_j. In our example, we have:

$$A = \begin{bmatrix} 0 & 1 & 0 & 0 & 1 & 0 \\ 1 & 0 & 1 & 0 & 1 & 0 \\ 0 & 1 & 0 & 1 & 0 & 0 \\ 0 & 0 & 1 & 0 & 1 & 1 \\ 1 & 1 & 0 & 1 & 0 & 0 \\ 0 & 0 & 0 & 1 & 0 & 0 \end{bmatrix}, \tag{5.1}$$

where the size of the system in $N = 6$. The number of edges can be computed directly from A as:

$$L = \frac{1}{2} \sum_{i=1}^{N} \sum_{j=1}^{N} a_{i,j}, \tag{5.2}$$

where the normalization factor $1/2$ is used because the network is undirected. By inspection, we can verify that $L = 7$.

Another important concept is the *degree of a node*, which indicates how many links to another node it has. Mathematically, the degree of node i is:

$$k_i = \sum_{j=1}^{N} a_{i,j} - a_{i,i}. \tag{5.3}$$

In our example, the nodes have the following degrees: $k_1 = 2$, $k_2 = 3$, $k_3 = 2$, $k_4 = 3$, $k_5 = 3$, and $k_6 = 1$. It is also interesting to note that as the size grows, a statistical analysis of the network based on probability theory becomes relevant to characterize its structure. We will present such an approach in the next section.

The links may also have some specific characteristics, such as space limitation (e.g. the capacity of the highway), the distance from one place to another (e.g. distance between the terminals in the airport), and strength of the relation (e.g. how many papers two researchers wrote together). In this case, the network is composed of *weighted links*. The elements of the matrix A are then not only zeros and ones, but they may assume numerical values that reflect the weight of the respective link. Let us consider a modification of the matrix presented in (5.1) considering now weighted links.

$$A_{\mathrm{w}} = \begin{bmatrix} 0 & 2 & 0 & 0 & 1 & 0 \\ 2 & 0 & 1 & 0 & 1 & 0 \\ 0 & 1 & 0 & 1 & 0 & 0 \\ 0 & 0 & 1 & 0 & 1 & 4 \\ 1 & 1 & 0 & 1 & 0 & 0 \\ 0 & 0 & 0 & 4 & 0 & 0 \end{bmatrix}, \tag{5.4}$$

where A_{w} is the adjacency matrix with weighted links.

In the case of researchers' network, this represents that researchers 4 and 6 have four papers together and researchers 1 and 2 have two; this indicates a stronger link between them. This may also indicate how many people can move from one building to another at the same time through that connection, or how far the airport terminals are.

Other important concept is the *path length*, which indicates how many links there are between two nodes. It is usual to have different paths connecting two nodes, and thus, their lengths also vary. The *distance* between two nodes is then defined as the *shortest path length* between them. The example presented in Figure 5.2 shows that from node 5 to node 6 there are two possible paths: $5 \rightarrow 4 \rightarrow 6$, or $5 \rightarrow 2 \rightarrow 3 \rightarrow 4 \rightarrow 6$. The first has a length of two while the second four. In this case, the distance between nodes 5 and 6, denoted $d_{5\rightarrow6}$, is given by $d_{5\rightarrow6} = \min(2,4) = 2$.

It is also possible to identify the number of shortest paths between two edges. In the previous case there is only one shortest path. If we now consider the paths between node 2 and node 6, we can show by inspection that there are two shortest paths, whose distance is $d_{2\rightarrow6} = 3$. It is also possible to compute them directly from the adjacency matrix as presented in Chapter 2 of [2]. The *diameter of the network* is defined as the maximum distance between two nodes. We can prove that the diameter of the network presented in Figure 5.2 is three.

There are other forms utilized to characterize networks. For example, a network is called *connected* if there is at least one path from any two nodes; otherwise the network is called *disconnected*. If all the nodes are connected to each other, we have a *complete network* or a *clique*. A *component* – also called *cluster* – is a subset of the network where at least one path from any two nodes exists. A link between two components is called *bridge*. The link between nodes 4 and 6 in Figure 5.2 is a bridge linking two components: one is the node 6 itself, and the component is the network formed by nodes 1, 2, 3, 4, and 5. If the bridge is broken, then the graph will become disconnected.

It is also possible to compute the *cluster coefficient* from any node in the network as follows:

$$C_i = \frac{2L_i}{x_i(x_i - 1)},\tag{5.5}$$

where L_i is the number of links between the x_i neighbors of node i. Note that the cluster coefficient is a number between 0 and 1 so that $C_i = 0$ means that the neighbors of node i are not linked to each other; $C_i = 1$ indicates that i and its neighbors form a complete graph. In our example, node 1 has $x_1 = 2$ (two neighbors: 2 and 5) and $L_1 = 1$ (one link between them), then $C_1 = 1$ indicating that the "subnetwork" is formed by 1, 2, and 5 is complete. Node 3 has $x_3 = 2$ (two neighbors: 2 and 4) and $L_3 = 0$ (no link between them), then $C_3 = 0$ indicating that 2 and 4 are not connected. Now, node 5 has $x_5 = 3$ (two neighbor: 1, 2 and 4) and $L_1 = 1$ (one link between 1 and 2), then $C_5 = 1/3$ indicating the existence of some connections between neighbors, whose level of connectedness is measured by the clustering coefficient.

Table 5.1 summarizes the main characteristics of a network. This list is not exhaustive and many others exist, while new ones may also be proposed. References [1, 2, 5] provide an in-depth analysis of networks and ways of characterizing them. An introduction to the mathematical formalism of graph theory can be found in [6]. In the next section, we will explore different types of networks.

Table 5.1 Summary of possible characteristics of a network G as defined in Definition 5.1.

Characteristic	Meaning	Formulation
Size	Number of nodes	N
Adjacency matrix	Representation of links	$A = (a_{i,j})$ with $a_{i,j} = 0$ or 1
Weighted adjacency matrix	Representation of links with weights	$A_{\mathrm{w}} = (a_{\mathrm{w};i,j}) \in \mathbb{R}^{N \times N}$
Node degree	Number of connections a given node i has	$k_i = \sum_{j=1}^{N} a_{i,j} - a_{i,i}$
Number of links	How many links exist in the network	Directed: $L = \sum_{i=1}^{N} \sum_{j=1}^{N} a_{i,j}$; Undirected: $L = \frac{1}{2} \sum_{i=1}^{N} \sum_{j=1}^{N} a_{i,j}$
Path length	Number of links from node i to node j; there can be more than one path	$l_{i \to j} \in \mathcal{L}_{i \to j}$, where $\mathcal{L}_{i \to j}$ contains all paths lengths from i to j
Distance	Shortest path length between two nodes	$d_{i \to j} = \min\limits_{l_{i \to j} \in \mathcal{L}_{i \to j}} l_{i \to j}$
Diameter of a network	Maximum distance between two nodes of the network	$\mathrm{diam}(G) = \max\limits_{i,j \in \mathcal{V}} d_{i \to j}$
Connectedness	Indicate if the network is connected, or disconnected	—
Clique	Indicate parts of the network where all nodes are connected, and thus, form a complete graph	—
Cluster or component	Part of a network where there is at least one path between any two nodes	—
Bridge	Connection between two components	—
Cluster coefficient	Indication of how connected a given node i is with respect to its neighbors	$C_i = \frac{2L_i}{x_i(x_i - 1)}$, where L_i is the number of links between the x_i neighbors of node i.

5.2 Network Types

In the previous section, we introduced some of the main metrics to characterize networks using a simple example presented in Figure 5.2. This is actually a static graph that represents a relatively small system. Networks can be (and usually are) much larger, possibly changing over time. Think about the Internet itself,

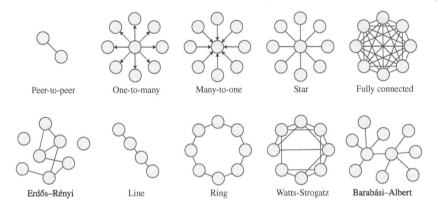

Figure 5.3 Example of different network topologies.

or social media applications. In this section, we will introduce different types of networks, from the simplest peer-to-peer network with only two elements to large random networks. Figure 5.3 presents an example of the network topologies to be discussed here.

5.2.1 Peer-to-Peer Networks

This is a topology that consists of two nodes that are connected to each other. In general, a peer-to-peer network with the size N is composed of $L = N/2$ links, where each node is only connected with another node. In this case, all nodes have degree one, i.e. $k_i = 1, \forall\, i \in \{1, \dots N\}$. This can be seen as an extreme case of a disconnected network. For example, the communication proposed by Shannon presented in the previous chapter with a transmitter and a receiver can be seen as a peer-to-peer network with $N = 2$ with one directed link.

5.2.2 One-to-Many, Many-to-One, and Star Networks

The **one-to-many** topology is a directed network where one "central" node is connected to all other nodes. If the network has size N, then we can assume without loss of generality that node 1 is the central one. Because the network is directed, the degree of the node can be categorized as out- and in-flows, and thus, $k_1^{out} = N - 1$ and $k_1^{in} = 0$. For all the other nodes, we have: $k_i^{out} = 0$ and $k_i^{in} = 1, \forall\, i \in \{2, \dots N\}$. The **many-to-one** topology can be understood as a mirrored version of the one-to-many. In this case, the central node have in-flow links from all the other nodes. Then, $k_1^{out} = 0$ and $k_1^{in} = N - 1$, while $k_i^{out} = 1$ and $k_i^{in} = 0, \forall\, i \in \{2, \dots N\}$. The **star** network has the same topology as the one-to-many and the many-to-one but with undirected links. In this case, we have $k_1 = N - 1$, while $k_i = 1, \forall\, i \in \{2, \dots N\}$. In all three cases, $L = N - 1$.

5.2.3 Complete and Erdös–Rényi Networks

A undirected network where all nodes are connected is called **complete** or **fully connected**. In a network with size N, $k_i = N - 1, \forall\, i \in \{1, \ldots N\}$ with a cluster coefficient $C_i = 1$. In this case there will be $L = N(N - 1)$ links in the network.

Erdös–Rényi (ER) networks were introduced by Hungarian mathematicians Erdös and Rényi in [7] to study random graphs with N nodes, which is a fixed number. The uncertainty in this network is related to the randomness of (i) the number of links, and (ii) the connection between nodes created by these links. Extreme cases are when the network has no links (i.e. it is only a collection of disconnected points) and a complete network where all the nodes are connected.

Without loss of generality, consider any two nodes i and j and a potential link $e_{i,j}$ between them. The existence of $e_{i,j}$ is associated with a random variable X with a sample space $S_X = \{0,1\}$ where $X = 0$ is associated with the absence of $e_{i,j}$ while $X = 1$ denotes the existence of such a link. The probability mass function of X is $p_X(0) = 1 - p$ and $p_X(1) = p$. In ER networks, the existence of a link between any two nodes in the network is determined by the outcomes of the independent outcomes of the random variable X.

In this case, a probabilistic characterization of the graph is possible. For example, each node in the network has $p(N - 1)$ links on average. If $p = 1$, the ER network is reduced to a complete graph. Likewise, if $p = 0$, the ER network is just a collection of points. They are the aforementioned extreme cases, which are actually deterministic. Note that the ER network is completely characterized by it size N and the probability p. A formalization of ER networks is the target of Exercise 5.1. A detailed mathematical analysis of ER networks can be found in [1, 2, 5, 7].

5.2.4 Line, Ring, and Regular Networks

The **line** network is defined when the nodes are connected to form a line, and thus, all nodes have two neighbors except the border nodes. If its size is N, then the network will have $L = N - 1$ links. If we assume that the border nodes are nodes 1 and N, then the node degrees are $k_1 = k_N = 1$ and $k_i = 2, \forall\, i \in \{2, \ldots, N - 1\}$. This network is connected and the cluster coefficient is $C_i = 0, \forall i \in \{1, \ldots, N\}$. The diameter of the network is $N - 1$, determined by the distance between the two extremes.

If the **ring** topology is similar but the border nodes are connected, then $L = N$ and $k_i = 2, \forall\, i \in \{1, \ldots, N\}$. If $N > 3$, then $C_i = 0, \forall i \in \{1, \ldots, N\}$. The diameter of this network considering both directed and undirected links is part of Exercise 5.2, and thus, we will not present it here.

Regular topologies are a broader term that usually refers to any topology that has a deterministic pattern. The already discussed line and ring topologies, as well as the star topology, can be classified as regular. Other regular grids are, for example, a rectangular, or a hexagonal grid, and also variations of the ring topology where nodes are symmetrically connected with m neighbors. Regular networks are deterministic and symmetric in contrast to random networks like ER.

5.2.5 Watts–Strogatz, Barabási–Albert and Other Networks

Watts–Strogatz (WS) networks are random graphs proposed by Watts and Strogatz in a paper published in 1998 [8]. The idea was to capture the small-world phenomena illustrated by the famous study by Milgram [9], which indicated that, even in very large social networks, the average path lengths between nodes are small. Watts and Strogatz proposed a method to generate networks with such a characteristic starting from a variation of a ring topology where nodes are connected with $m > 2$ neighbors. From the initial regular topology, nodes are randomly rewired with a given probability p. In the extreme cases, the WS network is a regular grid if $p = 0$, and a ER network is $p = 1$. The example presented by the authors is a pedagogical way to explain how a WS network can be generated [8]:

> Random rewiring procedure for interpolating between a regular ring lattice and a random network, without altering the number of vertices or edges in the graph. We start with a ring of N vertices, each connected to its m nearest neighbors by undirected edges. (For clarity, $N = 20$ and $m = 4$ in the schematic examples shown here, but much larger N and m are used in the rest of this Letter.) We choose a vertex and the edge that connects it to its nearest neighbor in a clockwise sense. With probability p, we reconnect this edge to a vertex chosen uniformly at random over the entire ring, with duplicate edges forbidden; otherwise we leave the edge in place. We repeat this process by moving clockwise around the ring, considering each vertex in turn until one lap is completed. Next, we consider the edges that connect vertices to their second-nearest neighbors clockwise. As before, we randomly rewire each of these edges with probability p, and continue this process, circulating around the ring and proceeding outward to more distant neighbors after each lap, until each edge in the original lattice has been considered once. (As there are $Nm/2$ edges in the entire graph, the rewiring process stops after $m/2$ laps.) Three realizations of this process are shown, for different values of p. For $p = 0$, the original ring is unchanged; as p increases, the graph becomes increasingly disordered until for $p = 1$, all edges are rewired randomly. One of our main results is that for intermediate values of p, the graph is a small-world network: highly clustered like a regular graph, yet with small characteristic path length, like a random graph.

This text refers to Figure 5.4, which is adapted from the original paper.

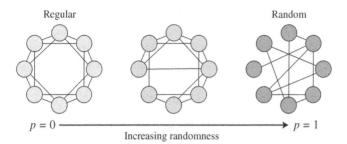

Regular   Random

$p = 0$ ⟶ $p = 1$

Increasing randomness

Figure 5.4 Example of a WS network. Source: Adapted from [8].

Barabási–Albert (BA) networks were proposed in [10] considering a scenario where new nodes are added over time by a probabilistic mechanism called *preferential attachment*. The idea behind this approach is that any new node has higher chances to be connected with the nodes that have higher degrees, which leads to the emergence of hubs (i.e. nodes that are highly connected).

In fact, BA networks are a particular case of *scale-free networks*, defined when the degree distribution of nodes follows a *power law probability distribution*. Mathematically, a power law distribution is defined as $P(k) \sim k^{-\gamma}$ with $\gamma > 1$, being then a heavy-tailed distribution (i.e. it is not exponentially bounded like Gaussian or Poisson distributions). Power laws have also the interesting property of scale invariance: for a given function $f(x) = cx^{-\gamma}$ with $c \in \mathbb{R}$ being an arbitrary constant and a scaling factor $a > 0$, we have $f(ax) = ca^{-\gamma}x^{-\gamma} = a^{-\gamma}f(x)$.

Although scale-free networks have received attention as a tool to model real-world interactions, recent results indicate that the scale-free property might not be as widespread as initially thought [11]. Besides, there are several other alternatives of networks such as tree, mesh, and bus, as well as hybrid topologies or those without a specific type.

Such a structural characterization is important when describing processes that happen *on* networks, such as transference of data packets on the Internet, possible flight routes between two cities, or propagation of viruses. This is the focus of the next section.

5.3 Processes on Networks and Applications

So far, we have shown how to characterize the structure of networks, also including procedures used to generate random graphs through rewiring as in WS or through preferential attachment as in BA. Networks are usually related to structural components where operating and flow components exist to jointly constitute a given system. In that case, networks represent the infrastructure that supports processes that constitute a functioning system.

The following subsections exemplify three different *processes on networks*, namely a data communication system, transportation in cities, and virus propagation. Our aim is to show how the network topology directly affects such processes in an intuitive manner without formulating the problem using the mathematical tools of dynamical systems.

5.3.1 Communication Systems

As indicated by Shannon's model [12], the peculiar function of a communication system is to transfer data from one point in space and/or time to another. Details of communication and computer networks can be found in textbooks, such as [13, 14]; detailed historical notes can be found in [15]. Consider two persons: Amanda and Bob. Amanda is willing to tell Bob that she will take a new course on cyber-physical systems.

If both Amanda and Bob are in the same room, she can just tell it in a peer-to-peer link. However, if they are in a lecture room far from each other, Amanda decides to write this on a piece of paper and send it to him through a path with two other nodes that relay the message. In this last case, the message is routed in a regular grid. If Amanda and Bob are in their own houses, Amanda decides to call Bob by a landline phone that is connected through a public switched telephone network (PSTN), which has a hierarchical tree-like topology with levels of switching systems. Amanda could also use her connection to the Internet either by sending an instant message from her computer, or by calling Bob by a Voice over Internet Protocol (VoIP) using her mobile phone connected to a cellular network of fourth generation (4G). The Internet forms a dynamic decentralized network with hubs, which could be characterized under some conditions as scale-free that may follow a BA network, e.g. [2].

Each real-world situation has characteristics of its own. For example, if Amanda is telling the news to Bob in the same room, there might be more people talking, or loud music, and thus, the peer-to-peer link may break. In this case, the two nodes are not connected and the network stops functioning. As a consequence, we can say that the network is not *robust* against this missing link. The situation described in the lecture room is different: the written message can reach its final destination through several paths, not necessarily the shortest one. If a link is broken because an absent node (i.e. no one is sitting at that table to forward the message to the destination), alternative routes may still be possible. We can then say that this network has a certain level of *resilience* with respect to missing links, i.e. even without a few links, the function of the communication network can be accomplished. However, if certain links are absent, the network may not percolate, and thus, the message cannot reach its final destination.

We could also compare the difference of calling by using PSTN and VoIP. Without going into technical details, PSTNs are based on a dedicated hierarchical network that resembles a star topology, and thereby, with important central nodes that connect Amanda and Bob by multiple circuit switches. The robustness of the network as a whole is highly dependent on its central nodes. Once established, this link is exclusive with quality guarantees. VoIP, in its turn, runs on the Internet, which is a dynamic network-based data multiplexing using packet switching technology, without explicit quality guarantees. VoIP "slices" the voice into smaller data packets that travel through the network in different routes to reach the address of the final destination, multiplexed with all other possible data packets (e.g. other VoIP, videos, and text files). At the destination, the packets are grouped together and decoded as voice. The quality of a VoIP call is usually worse than that of PSTN calls, but the network is more resilient because of the existence of several hubs and many possible routes.

5.3.2 Transportation in Cities

The case studied by Euler presented in Figure 5.1 can be seen as an example of how transportation networks could be modeled as a graph. This was clearly a toy example, but very illustrative because it indicates the impossibility of a solution to the Seven Bridges problem. One of the classical problems in the literature of transportation networks is the *traveling salesman problem* [16]. The problem is the following: *Given a list of cities with their connections, what is the shortest possible route that visits each city exactly once and returns to the origin city?* This problem can be formulated as a graph with weighed edges, which results in a combinatorial optimization problem; in the literature of computer sciences, this is known as a typical example of a problem that cannot be solved in polynomial time.

Without going into the details of NP-hardness and its importance for operational research and logistics, we will focus on topological aspects of the network and how this affects transportation in cities. Consider the Berlin S-Train map presented in Figure 5.5. Each station is a node connected by edges, which are associated with different train lines. Outer stations have a low degree while central stations high, resembling a star topology. Around the center of the network there are also lines forming a ring topology. Putting together, this hybrid topology is built to have not only one overcrowded central hub, but also a few smaller hubs where the radial lines cross with the ring. In this way, the traffic of the transportation network could be balanced between different routes.

This simple daily life example illustrates how networks are part of our lives. It is interesting to mention that studying cities through the lenses of network sciences

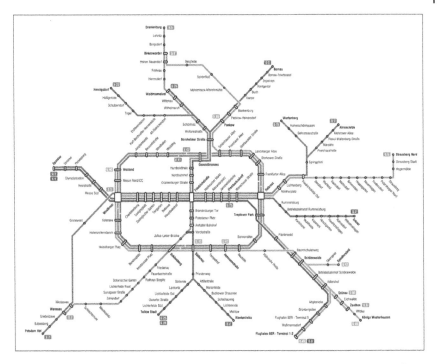

Figure 5.5   Map of the Berlin S-Train. Source: https://commons.wikimedia.org/wiki/File: S-Bahn_Berlin_-_Netzplan.svg.

has been a tendency for a few years. For interested readers, the reference [17] is a milestone in the field by empirically showing how different characteristics of cities scale with their size along with a formalism derived from network sciences and dynamical systems.

5.3.3   Virus Propagation and Epidemiology

The COVID-19 pandemic has led to a widespread interest in epidemiological models of how viruses and other infectious deceases propagate. In the literature there are three basic models [2], namely Susceptible-Infected (SI), Susceptible-Infected-Susceptible (SIS), and Susceptible-Infected-Recovered (SIR). The premise employed in those models is that individuals can be classified with respect to a given pathogen as:

- **Susceptible:** Healthy individuals that have not yet contacted the pathogen;
- **Infected:** Individuals that carry the pathogen and can infect susceptible individuals if they are in contact;

- **Recovered:** Individuals who have been infected before, but are healthy, and thus, cannot propagate the pathogen.

Without going into the mathematical characterization based on differential equations and random walking theory, the idea of SI, SIS, and SIR basic models is to analytically assess when a disease will be spread over all population, or the effect of the spreading rate – the famous R_0 that dominated the news in 2020 during the COVID-19 pandemic, which, in the words of Barabasi [2], is *the number of new infections each infected individual causes under ideal circumstances.*

The SI model is the one that assumes that there is no recovery, and thus, it indicates how long a given population will be fully composed of infected individuals. The SIS model assumes that individuals can recover, but they are immune and they can become infected once again. In this case, two outcomes at the population level may happen: (i) an endemic state where there is an equilibrium so that there is always a given ratio of the population infected, or (ii) a decease-free state where the pathogen cannot propagate and fades away from the population. The situation (i) is associated with $R_0 > 1$, while (ii) with $R_0 < 1$. The SIR model adds another class of individuals that are not susceptible to the pathogen and also do not transmit it. These are recovered individuals, who are immune to the pathogen after infection. In this basic SIR model, all the elements of the population will be eventually recovered. Those models could also be adapted to include immunization and individuals that died.

Once again, it is interesting to quote Barabasi [2]:

> In summary, depending on the characteristics of a pathogen, we need different models to capture the dynamics of an epidemic outbreak. (…) [T]he predictions of the SI, SIS, and SIR models agree with each other in the early stages of an epidemic: When the number of infected individuals is small, the disease spreads freely and the number of infected individuals increases exponentially. The outcomes are different for large times: In the SI model everyone becomes infected; the SIS model either reaches an endemic state, in which a finite fraction of individuals are always infected, or the infection dies out; in the SIR model everyone recovers at the end. The reproductive number predicts the long-term fate of an epidemic: for $R_0 < 1$ the pathogen persists in the population, while for $R_0 > 1$ it dies out naturally.

For this reason, we can see the importance of the R_0 factor so highly discussed during the COVID-19 pandemic.

In a seminal paper [18], Pastor-Satorras and Vespignani expanded those basic epidemiological models from generic dynamical equations to scenarios with a networked structure. In this case, individuals are nodes in the network whose links refer to pairwise contacts between individuals. The propagation of a pathogen is actually a result of social interactions, including transportation networks in cities and daily activities, as well as international traveling and popular sports or music events. As indicated by studies of cities and transportation networks, epidemiological networks are usually associated with scaling laws, and thus, there will be some nodes with a very high degree. Such nodes are *super-spreaders*. Although each pathogen is different, understanding the networked dynamics of its propagation is essential to design proper interventions to combat epidemics [2, 19].

Figure 5.6 illustrates the propagation of different flu strains using the open-source tool called *Nexttrain* [20]. It is interesting to see how different strains propagate as a network connecting different regions. The granularity of this tool is not enough to capture the networked dynamics of the SI, SIS, and SIR models, but it illustrates the effects of international traveling in flu epidemics. Note that an additional aspect can be added in the model to include exposures, defined as the immediate neighbors of a given infected node. This model is known as Susceptible-Exposed-Infected-Susceptible (SEIR) [21].

5.4 Limitations

Understanding networks and processes on networks is highly important. The year of 2020 serves as an illustrative example: networked models have been developed to combat the COVID-19 pandemic and the propagation of fake news in social media. The main limitations of network sciences are related to topics already discussed in Chapter 1: the relation between a particular scientific theory and its respective object. Despite being a controversial topic, we think it is important to indicate the dangers of either data-driven or purely mathematical models, as well as a tendency of universality that is dominant in the field.

5.4.1 From (Big) Data to Mathematical Abstractions

Big data and data-driven approaches where "machines learn" patterns from data are widespread. As indicated by *Nextstrain* and [2], empirical data can be used to build mathematical models capable of predicting events such an outbreak. However, as reported in [11], modeling (big) data related to complex networks is not straightforward and may create a tendency of empirically finding scale-free networks. Such an approach may lead to empirical scaling laws that may lead to inaccurate predictions. Besides, if the predictive model affects the dynamics of the network evolution, the mathematical abstraction may become intrinsically incorrect because of a feedback loop.

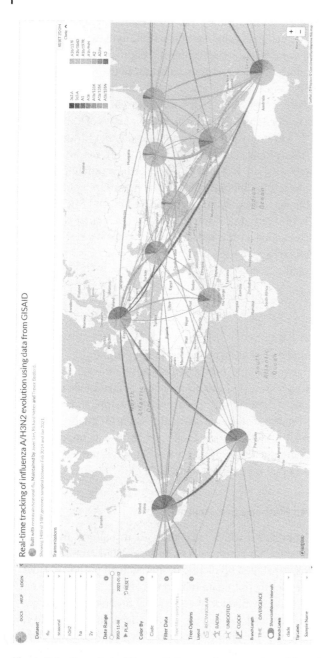

Figure 5.6 Snapshot of the spatiotemporal propagation of flu using *Nextstrain*. Generated through https://nextstrain.org/flu/.

5.4.2 From Mathematical Abstractions to Models of Physical Processes

Generative mathematical models that assume an ideal, purified object to characterize complex real-world networks may also be problematic. Similar to what we have described before about empirical data being "forced" to fit the characterization of scale-free networks, generative mathematical models for networks may likewise be misleading. For instance, predictions for COVID-19 based on generative models can be *wrong but useful* as discussed in [22]. What should be highlighted is that mathematical models cannot impose itself on the reality of the process under consideration. If the data obtained from the real world network do not behave as expected by the generative model, the evaluation ought to be fair and not blame the reality for not being the ideal model. As indicated in Chapter 1, a consistent and elegant mathematical model is neither a necessary nor a sufficient condition for being a scientific theory of a given object, including networks.

5.4.3 Universality and Cross-Domain Issues

The two limitations just presented can, in fact, be summarized in one problem: universality. This is closely related to the discussions introduced by Broido and Clauset [11] in contrast to the view that scale-free networks are everywhere, defended by Barabasi [2]. The question was an interesting one posed by Holme [23], as a comment about such a controversy. Similar questions might be considered by the science of cities defended by Bettencourt [17] utilizing network sciences to derive the characteristics of complex urban formations.

The question is the following: are power grids, cities, epidemics, social relations, and air traffic all following a hidden pattern revealed by network sciences? This is controversial and, for the reasons stated in Chapter 1, the position taken by this book leads to a no as the answer, although the models are indeed useful if understood a tool, not as a scientific representation of reality. Note that some objects can be scientifically characterized as networks, but it is not a universal characterization. Most of the time, network sciences provides a potentially useful model, not a scientific theory of the phenomenon under investigation.

Inga Holmdahl and Caroline Buckee precisely indicate in [22] five questions about models for COVID-19, also stating how their results should be used. Such precautionary propositions touch the key issues mentioned above by identifying two classes of models, namely data-driven predictive models and mechanistic mathematical models. This is what they wrote:

FIVE QUESTIONS TO ASK ABOUT MODEL RESULTS

1) What is the purpose and time frame of this model? For example, is it a purely statistical model intended to provide short-term forecasts or a mechanistic model investigating future scenarios? These two types of models have different limitations.

2) What are the basic model assumptions? What is being assumed about immunity and asymptomatic transmission, for example? How are contact parameters included?

3) How is uncertainty being displayed? For statistical models, how are confidence intervals calculated and displayed? Uncertainty should increase as we move into the future. For mechanistic models, what parameters are being varied? Reliable modeling descriptions will usually include a table of parameter ranges – check to see whether those ranges make sense.

4) If the model is fitted to data, which data are used? Models fitted to confirmed Covid-19 cases are unlikely to be reliable. Models fitted to hospitalization or death data may be more reliable, but their reliability will depend on the setting.

5) Is the model general, or does it reflect a particular context? If the latter, is the spatial scale – national, regional, or local – appropriate for the modeling questions being asked and are the assumptions relevant for the setting? Population density will play an important role in determining model appropriateness, for example, and contact-rate parameters are likely to be context-specific.

(…)

Unlike other scientific efforts, in which researchers continuously refine methods and collectively attempt to approach a truth about the world, epidemiologic models are often designed to help us systematically examine the implications of various assumptions about a highly nonlinear process that is hard to predict using only intuition. Models are constrained by what we know and what we assume, but used appropriately and with an understanding of these limitations, they can and should help guide us through this pandemic.

We think that the argument put forth by the authors is also valid across several other domains in general and network sciences in particular, also including the aforementioned debate about whether scale-free networks are *rare or everywhere* [23].

5.5 Summary

This chapter presented an overview of the most fundamental concepts of graph theory and network sciences. The aim was to introduce the formalism used to model relations between different elements, being then material or symbolic. We presented structural characteristics of networks and also processes on networks, both illustrated by examples from our daily lives. The limitations of network models were also discussed focusing on the dangers of either data-driven or mathematical models. Interested readers could find a good introduction to graph theory in [6]. The core concepts and methods in network sciences are presented in an interactive manner in [2]. More advanced discussions about complex networks are presented in [5]. Current debate about epidemiological models and scale-free networks are presented in [22] and [23], respectively.

Exercises

5.1 Formalization of ER networks. ER networks are presented in Section 5.2.3. Consider a network with N nodes and a probability p that a link between two nodes exists.
  (a) If Y represents a random variable associated with the degree of a given node, compute the probability mass function $P(Y = k)$, i.e. the probability that a given node in the network has k connections.
  (b) Compute the expected value of Y, i.e. $E[Y]$.
  (c) Compute the variance of Y, i.e. $var(Y) = \mathbb{E}\left[Y^2\right] - (E[Y])^2$
  (d) Interpret such theoretical results considering two scenarios: (i) N varies from two to infinity, and (ii) p varies from zero to one. Write this answer with your own words supported by the mathematical findings.

5.2 Network topologies. Consider a network with a ring topology of $N \geq 4$ nodes with links that are unweighted and undirected. Present the results as a function of N.
  (a) Compute the degrees of the nodes.
  (b) Present the adjacency matrix.
  (c) Compute the network diameter.
  (d) Compute the cluster coefficients of the nodes.
  (e) Analyze the ring topology with $N = 20$ as a communication network (i.e. how data travel to a point to another in the network) based on the node degree, the network diameter, and the cluster coefficient.
  (f) Consider that you are a designer who wishes to improve the communication network described in (e). Your task is to change the ring network

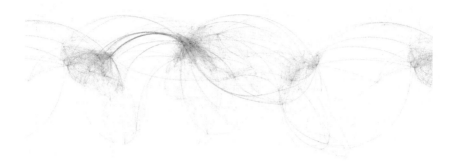

Figure 5.7 World airline routes. Source: https://commons.wikimedia.org/wiki/File:
World_airline_routes.png.

topology by adding only one new node and an unlimited number of links
connected to it. Justify your decision and identify a vulnerability of this
topology.

5.3 Networks in the real world. Consider the world airline routes presented
in Figure 5.7.
(a) Provide a qualitative assessment of this network by, for instance, com-
menting about its topology and node degree distribution.
(b) How many clusters could you visually identify in this network?
(c) Consider that a new infectious disease transmitted by airborne droplet
like flu appeared in the world. Which continent would be easiest to isolate?
Justify.

References

1 Newman ME. The structure and function of complex networks. SIAM Review.
2003;45(2):167–256.

2 Barabási AL. Network Science. Cambridge University Press; 2016. Available
online: http://networksciencebook.com/.

3 Euler L. The seven bridges of Königsberg. The World of Mathematics.
1956;1:573–580. Original version available online: http://eulerarchive.maa
.org//docs/originals/E053.pdf.

4 Wikipedia. Seven Bridges of Königsberg—Wikipedia, The Free Encyclopedia;
2021. [Online; accessed 3-February-2021]. Available from: https://en.wikipedia
.org/wiki/Seven_Bridges_of_K%C3%B6nigsberg.

5 Thurner S, Hanel R, Klimek P. Introduction to the Theory of Complex Systems. Oxford University Press; 2018.

6 Diestel R. Graph Theory. Springer; 2017.

7 Erdős P, Rényi A. On the evolution of random graphs. Publication of the Mathematical Institute of the Hungarian Academy of Sciences. 1960;5(1):17–60. Original version available online: https://www.renyi.hu/p_erdos/1959-11.pdf.

8 Watts DJ, Strogatz SH. Collective dynamics of 'small-world' networks. Nature. 1998;393(6684):440–442.

9 Milgram S. The small world problem. Psychology Today. 1967;2(1):60–67.

10 Barabási AL, Albert R. Emergence of scaling in random networks. Science. 1999;286(5439):509–512.

11 Broido AD, Clauset A. Scale-free networks are rare. Nature Communications. 2019;10(1):1–10.

12 Shannon CE. A mathematical theory of communication. The Bell System Technical Journal. 1948;27(3):379–423.

13 Kurose JF, Ross KW. Computer Networking: A Top-Down Approach. Pearson; 2016.

14 Popovski P. Wireless Connectivity: An Intuitive and Fundamental Guide. John Wiley & Sons; 2020.

15 Huurdeman AA. The Worldwide History of Telecommunications. Wiley Online Library; 2003.

16 Flood MM. The traveling-salesman problem. Operations Research. 1956;4(1):61–75.

17 Bettencourt LM. The origins of scaling in cities. Science. 2013;340(6139):1438–1441.

18 Pastor-Satorras R, Vespignani A. Epidemic spreading in scale-free networks. Physical Review Letters. 2001;86(14):3200.

19 Althouse BM, Wenger EA, Miller JC, Scarpino SV, Allard A, Hébert-Dufresne L, et al. Superspreading events in the transmission dynamics of SARS-CoV-2: opportunities for interventions and control. PLoS Biology. 2020;18(11):e3000897.

20 Hadfield J, Megill C, Bell SM, Huddleston J, Potter B, Callender C, et al. Nextstrain: real-time tracking of pathogen evolution. Bioinformatics. 2018;34(23):4121–4123. Tool available: https://nextstrain.org/.

21 He S, Peng Y, Sun K. SEIR modeling of the COVID-19 and its dynamics. Non-linear Dynamics. 2020;101(3):1667–1680.

22 Holmdahl I, Buckee C. Wrong but useful-what covid-19 epidemio-logic models can and cannot tell us. New England Journal of Medicine. 2020;383(4):303–305.

23 Holme P. Rare and everywhere: perspectives on scale-free networks. Nature Communications. 2019;10(1):1–3.

6

Decisions and Actions

Decision-making processes are widespread and usually associated with actions as a means of achieving specific objectives. This could be the determination of operating points of industrial machines to maximize their outputs, the selection of the best route in the traveling salesman problem introduced in Chapter 5, or simply the decision to drink coffee before sleeping. This chapter will introduce different approaches to making decisions, being them centralized, decentralized, or distributed, and which are employed to govern actions based on informative data. In particular, we will discuss three different methodologies, namely optimization, game theory, and rule-based decision. Our focus will be on mathematical and computational methods; examples from humans or animals are presented only as pedagogical illustrations, and thus, we ought to proceed with great care to avoid extrapolating such results. Therefore, one important remark before we start: decision-making processes are generally normative, and thus, doctrinaire at some level even when they appear to be *natural* or *spontaneous*.

6.1 Introduction

Decision is defined in [1] as *the act or process of deciding* or *a determination arrived at after consideration*. However, decisions do not exist in the void: they always exist as a mediator between data and action. Hence, decision-making process refers to the way that decisions are made about possible action(s) given certain data as inputs. Decisions and actions are then close to each other, but they are different processes.

In this book, we assume that decisions always refer to potential actions, but that deciding and acting elements are not necessarily the same. In the following, we will define the main terminology used here to avoid misunderstanding.

Cyber-physical Systems: Theory, Methodology, and Applications, First Edition. Pedro H. J. Nardelli.

Definition 6.1   *Decision-making, decision-makers, and agents*   *Decision-making* is a process that uses data to select a given action, or a set of actions, among possible alternatives. *Decision-makers* are the elements that participate in decision-making processes based on logical relations. *Agents* are the elements capable of acting in accordance with the outcome of the decision-making process, and thus, they are symbolically controlled by the decision-makers' commands (and not by their physical force). The attributes of deciding and acting are not mutually exclusive, leading to elements that are (i) only decision-makers, (ii) only agents, or (iii) both.

This definition reaffirms that the main elements in decision-making processes are decision-makers, not agents. A mathematical formalism is possible if some specific conditions concerning their possible choices hold. This is the focus of decision theory [2], employed in behavioral psychology and neoclassical economics, and whose basic assumptions about human beings are frequently questioned (e.g. [3–5]). We also share those concerns, and thus, our conceptualization of decision-making will be presented in a form different from the ones usually found in the literature. Similar to previous chapters, we will introduce fundamental concepts and tools for decision-making, as well as their limitations. The first step is to identify the forms that decision-making processes can take.

6.2   Forms of Decision-Making

One way to classify decision-making processes is the topological characteristic of the network composed of decision-makers and agents connected through logical links. We can then discriminate decision-making processes into three types: centralized, distributed, and decentralized. Figure 6.1 provides an illustration of the different types.

In centralized decision-making, there is one element – the only decision-maker – that has the attribute of deciding on the actions taken by all agents in a given system or process. This approach usually assumes that the agents have no autonomy and just follow the prescribed actions determined by the decision-maker, and thus, agents are not involved in decision-making processes. The only exception is a special case in which the decision-maker itself is an agent – but a special agent because it determines the actions to be accomplished by itself and by all other agents. The network topology is usually the one-to-many topology, in which the central node is the decision-maker and the agents are at the edge.

In distributed decision-making, all elements are both potential decision-makers and potential agents, i.e. all have the capability of deciding and acting. In comparison with the centralized approach, distributed decision-making presumes

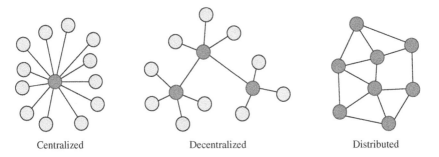

Figure 6.1 Examples of centralized, decentralized, and distributed networks.

that agents are relatively autonomous because they are part of decision-making processes that define the actions to be taken. Distributed decision-making is, in this sense, participatory and has two extreme regimes, namely: (i) unconstrained decision and action by all elements (i.e. full autonomy given the possible choices) and (ii) decision by consensus (i.e, all elements agree with the decision to be taken on individual actions). In between, there is a spectrum of possibilities, such as decision by the simple majority or by groups. Random networks with undirected links provide a usual topology for distributed decision-making.

Decentralized decision-making can be seen as a hybrid between the centralized and distributed cases, which consists of more than one decision-maker and may also include elements that are (i) only decision-makers, (ii) only agents, and (iii) both decision-makers and agents. In this case, decentralized decision-making may take different forms, some resembling a centralized structure (e.g. few decision-makers determining the action of many agents), others a distributed one (e.g. most of elements are both decision-makers and agents). It is also possible to have hierarchical decision-making, where some decision-makers of higher ranking impose constraints and rules on lower-ranked ones. Trees and networks with hubs are typical topologies of this case, although not the only ones.

In summary, the proposed classification refers to how the decision-makers are structured in relation to the agents. If there is only one decision-maker that directly commands and controls all elements, then the decision is centralized. If all the elements are both potential decision-makers and agents, then the decision is distributed. If there are more than one decision-maker coexisting with other elements, then the decision is decentralized.

It is important to reinforce that, in all three types, the autonomy of agents and decision-makers is constrained by the norms that govern the decisions and the possible choices for actions. Because both norms and choices are generally predefined, fixed and given, decision-making processes are doctrinaire. One important remark is that distributed decision-making is very often associated with autonomy for deciding and then acting. This view may be misleading because the level of autonomy is relative in the sense that decision-makers are always limited by the norms, and agents are always limited by the possible choices of action. The following example provides an illustration of different approaches of decision-making using an ordinary decision-making process of our daily lives.

Example 6.1 *Deciding on clothes for children* Children are usually stubborn, and thus, parents may have some hard time. One typical example is the selection of clothes that the children need to use to go to a birthday party. Assume a family with two parents and four children. This case could be framed as a decision-making process whose solutions could take the centralized, distributed, or decentralized forms as listed in the following.

- **Centralized:** One parent selects the clothes for all four children. This parent is the single decision-maker and the children the agents that will dress up.
- **Distributed:** The parents let the children select the clothes. Three children discuss with each other and find a consensus on using clothes with the same color; one child decides alone. In this case, the parents are not decision-makers, and the children are both decision-makers and agents.
- **Decentralized:** Both parents and the four children jointly deliberate about the clothes of each child. Another option is that both parents agree on which clothes the children will use but without asking them. In both cases, the parents are only decision-makers. However, the first case is closer to distributed decision-making because the children are both decision-makers and agents, while the second case is closer to the centralized one because the children are only agents but the final decision is distributed between the two parents.

Note, however, that the autonomy in all three cases is limited to the existing option of clothes. The decision on going to the party is excluded because it has been imposed by the parents beforehand.

This example is provided as a didactic illustration of the different types of decision-making processes. Although the proposed nomenclature is not standardized in the literature, the distinction between centralized, distributed, and decentralized decision-making is broadly used to indicate the tendency of economical and political decentralization in the post-Fordist age [6], which is also emulated by recent computer network architectures [7–9]. Another important reminder is that the relation between decision-makers and agents is necessarily mediated by data processes, and thus, the decision-makers logically control the agents by data commands, not by physical causation (which, in Example 6.1, could be the case that the parents are both decision-makers and agents who decide the clothes and directly dress the children). The following proposition generalizes this by using the concept of the level of processes introduced in Chapter 4.

Proposition 6.1 *Level of actions and decision-making processes* Action is always a physical process, and thus, agents are associated with level 0 processes. Decision-making is always a symbolic process, and thus, associated with levels greater than 0 so that: level 1 decision-makers have direct control of the agents' actions, level 2 decision-makers have direct control of the choices that level 1 decision-makers have, and so on. Regardless of the level of the decision-making, the set of all possible choices is grounded by physical actions, which are their ultimate limiting factor. Therefore, autonomy in decision-making and actions is always limited by the physical reality.

This proposition has interesting implications that will be explored throughout this book. At this point, though, we will slightly shift the focus and investigate different formal frameworks for decision-making processes as such, without considering the material reality they refer to.

6.3 Optimization

Optimization is a branch of mathematics that focuses on the selection of the optimal choice with respect to a given criterion among a (finite or infinite) set of existing alternatives. An optimization problem can be roughly defined by the questions: *what is the maximum/minimum value that a given function can assume and how to achieve it given a set of constraints?* Mathematically, we have [10]:

$$
\begin{aligned}
\min_{x} \quad & f_0(x) \\
\text{s.t.} \quad & f_i(x) \le b_i
\end{aligned}
\tag{6.1}
$$

where x is the *optimization variable*, $f_0(x)$ is the *objective function*, $f_i(x)$'s are the *constraint functions*, and the constants b_i's are the *constraint limits* for $i = 1, \dots, m$. The solution of (6.1), denoted by x^*, is called *optimal*, indicating that $f_0(x^*) \le f_0(y)$ for any z that satisfies the constraints $f_i(z) \le b_i$. The proposed optimization problem is valid for vectors, and thus, the optimization variable $x \in \mathbb{R}^n$, and the functions $f : \mathbb{R}^n \to \mathbb{R}$. Besides, a maximization problem can always be formulated as a minimization: $\max f_0(x) = \min -f_0(x)$. The optimization problem can be read as *what is the minimum value of f_0 in terms of x subject to the constraints defined by f_i and b_i?*

Mathematical optimization is clearly an important tool for decision-making regardless of its type. In the following, we will provide some examples to illustrate this.

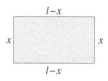
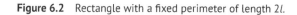

Figure 6.2 Rectangle with a fixed perimeter of length $2l$.

Example 6.2 *The largest rectangular area for a fixed perimeter* Consider a case that a decision-maker needs to choose the length of the two sides of a rectangle assuming that the perimeter is fixed and has a length of $2l$. If one side of the rectangle has a length x, the perpendicular sides have a length $l - x$. Figure 6.2 illustrates this. The area is then $x(l - x) = lx - x^2$.

We can then write the optimization problem:

$$\begin{aligned} \max_{x} \quad & lx - x^2 \\ \text{s.t.} \quad & x > 0 \\ & x < l \end{aligned} \quad\quad (6.2)$$

There are a few possible ways to solve the problem. One is to prove that the function $lx - x^2$ is concave in terms of x for the region of interest $0 < x < l$, and thus, find the maximum as:

$$\frac{d}{dx}(lx - x^2)\Big|_{x=x^*} = 0 \Rightarrow l - 2x^* = 0 \Rightarrow x^* = \frac{l}{2}.$$

As a result, the area is maximized but the rectangle being symmetric; therefore, a square of size $l/2$ and an area of l^2. For example, if the perimeter is $2l = 20$ m, then the solution to the problem is $x^* = 5$ m leading to an area of $25\,\text{m}^2$, as illustrated in Figure 6.3.

Example 6.3 *Random access in wireless networks* One of the main challenges in wireless communications is how transmitters could share the same medium without interfering with the reception of each other [11]. Consider the scenario depicted in Figure 6.4, where there are N transmitters trying to transmit to a given gateway in the center of a circle with an area A, forming a many-to-one network.

If we consider a fixed area A and transmitters that are mobile, N can be modeled as a random variable with expected value λA, and thus, λ represents the expected spatial density of transmitters. If all the transmitters try to access the shared medium, collisions become more likely. On the other hand, if not all transmitters are allowed to communicate, it is expected that the transmissions have a better chance to be successful. To control the access to the shared medium, each transmitter in the network has a random number generator that is designed to guide the individual decisions so that each will access the medium with a given probability p, which is equal for all transmitters.

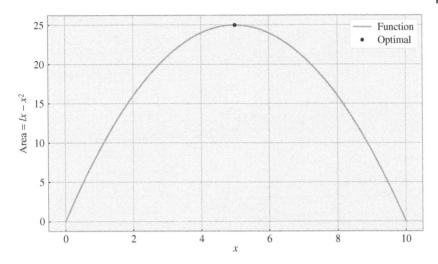

Figure 6.3 Area of a rectangle considering a fixed perimeter of 20 m. The maximum area is 25 m² achieved when all the sides have the same length of 5 m.

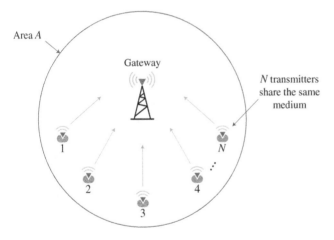

Figure 6.4 Random access in wireless networks.

This trade-off can be computed by a metric called spatial throughput S, which is the average density of transmitters times the probability that the individual transmissions are correctly decoded. As a rough estimation, we have $S = p\lambda \exp(-kp\lambda)$, where k is a constant that incorporates different factors related to the communication channel. In this case, there are two decision-making processes: (i) the definition of the probability p^* that maximizes S and (ii) individual transmitters that are both decision-makers and agents that will decide if they transmit in a

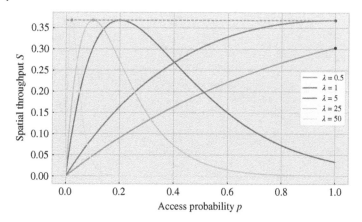

Figure 6.5 Spatial throughput in wireless networks as a function of the probability that a node accesses the communication channel.

given time slot with such a probability p^*. The case (i) is an optimization problem, while case (ii) is a rule-based decision, which will be discussed later on. The spatial throughput optimization problem can be formulated as follows:

$$\begin{aligned} \max_{p} \quad & p\lambda \exp(-kp\lambda) \\ \text{s.t.} \quad & p \leq 0 \\ & p \geq 1 \end{aligned} \qquad (6.3)$$

Solving the problem similarly to the previous example, we have $p^* = \min\left[1, (k\lambda)^{-1}\right]$, and then

- if $p^* = 1$, then $S^* = \lambda \exp(-k\lambda)$,
- if $p^* = (k\lambda)^{-1}$, then $S^* = k^{-1} \exp(-1)$.

The solution is exemplified in Figure 6.5, considering different values of λ and $k = 1$. Note that an area that is less populated by transmitters is exemplified by $\lambda = 0.5$ cannot achieve the maximum spatial throughput; however, the transmitters do not need any contention policy (i.e. always transmit with $p^* = 1$). The case with $\lambda = 1$ is special; the highest spatial throughput can be achieved without contention. However, for denser areas, the contention is needed to optimize the spatial throughput, leading to $p^* < 1$. Nevertheless, the maximum spatial throughput $S^* = \exp(-1)$ is always achieved if $\lambda \geq 1$.

These two examples illustrate how optimization is usually carried out in a pedagogical manner. Optimization problems can be much more complicated as indicated by the cases presented in [10]. Nevertheless, the formulation stated in (6.1) is generally valid for all optimization problems. Example 6.2 presented a case that the optimal solution is guaranteed by the problem statement. Example 6.3 is different because it *assumes* that all the nodes act in the same way employing the same access probability p while the optimization refers to a network-level metric called spatial throughput. If the goal is to maximize the link throughput, the optimal solution taken by individual transmitters would lead to different results as presented in [12] based on a more sophisticated scenario. The comparison between network and individual level optimization approaches will be the focus of Exercise 6.1. Note that in Example 6.3 the decision-making is centralized by an element that indicates to the agents the probability p^* that they need to use to access the radio resource. In this case, this central element is a level 2 decision-maker, and the transmitters are level 1 decision-makers and also agents.

In the next section, we will look at a different problem defined when decision-makers interact in a competitive manner, taking decisions based on their own self-interest.

6.4 Game Theory

Game theory is a mathematical theory used as a support for strategic decision-making in scenarios with more than one decision-maker. Before we start to describe its fundamentals, it is important to state different views of such a theory to avoid possible pitfalls.

- [13]: *Game theory is the mathematical study of interaction among independent, self-interested agents. The audience for game theory has grown dramatically in recent years, and now spans disciplines as diverse as political science, biology, psychology, economics, linguistics, sociology, and computer science, among others.*
- [14]: *Game theory is an axiomatic-mathematical theory that presents a set of axioms that people have to "satisfy" by definition to count as "rational." This makes for "rigorous" and "precise" conclusions – but never about the real world. Game theory does not give us any information at all about the real world. Instead of confronting the theory with real-world phenomena it becomes a simple matter of definition if real-world phenomena are to count as signs of 'rationality.' It gives us absolutely irrefutable knowledge–but only since the knowledge is purely definitional.*
- [15]: *In many respects this enthusiasm is not difficult to understand. Game theory was probably born with the publication of The Theory of Games and Economic Behaviour by John von Neumann and Oskar Morgenstern (first published in 1944 with second and third editions in 1947 and 1953). They defined a game as any interaction between agents that is governed by a set of rules specifying the possible moves for each participant and a set of outcomes for each possible combination*

of moves. One is hard put to find an example of social phenomenon that cannot be so described. Thus a theory of games promises to apply to almost any social interaction where individuals have some understanding of how the outcome for one is affected not only by his or her own actions but also by the actions of others. This is quite extraordinary. From crossing the road in traffic, to decisions to disarm, raise prices, give to charity, join a union, produce a commodity, have children, and so on, it seems we will now be able to draw on a single mode of analysis: the theory of games.

At the outset, we should make clear that we doubt such a claim is warranted. This is a critical guide to game theory. Make no mistake though, we enjoy game theory and have spent many hours pondering its various twists and turns. Indeed it has helped us on many issues. However, we believe that this is predominantly how game theory makes a contribution. It is useful mainly because it helps clarify some fundamental issues and debates in social science, for instance those within and around the political theory of liberal individualism. In this sense, we believe the contribution of game theory to be largely pedagogical. Such contributions are not to be sneezed at.

This book presents a pragmatical account of game theory as a model for decision-making under very specific assumptions, and thus, it is not considered to have any scientific value in terms of explaining human behavior, social dynamics, or nature. Game theory is an axiomatic construct that consists of decision-makers that have choices that will result in different payoffs assuming that they (i) are *instrumentally rational*, i.e. always aim at maximizing their individual payoffs, (ii) have background knowledge about each others' rationality but do not communicate to coordinate decisions, and (iii) are completely aware of the rules of the game. In the following, we will formalize this by defining games in normal form.

Definition 6.2 *Games in normal form and basic lexicon* Games in normal form are defined by:

- A set of decision-makers (called players): $D = \{1, \dots, k\}$.
- Each decision-maker $i \in D$ has a set of possible decisions about actions denoted A_i, which might be either finite or infinite. The selected decision $a_i \in A_i$ is called pure strategy.
- $A = A_1 \times \cdots \times A_k$ is the set of all profiles of pure strategies with a generic element defined by the vector $\mathbf{a} = (a_1, \dots, a_k)$, which denotes a profile of strategies.
- Decision-maker i payoff $u_i(\mathbf{a})$ is a function of the vector of actions $\mathbf{a}$ profile defined by the function $u_i : A \to \mathbb{R}$.

A strategy a_i is called *dominant* for a given decision-maker i if it leads to the maximum payoff $u_i^*(\mathbf{a})$ regardless of the decision of the other players. *Pure strategy Nash equilibrium* is the situation defined when the profile of strategies leads to the best response for all decision-makers, and thus, there is an equilibrium where individual deviations from the equilibrium strategy never produce higher payoffs.

The game with two decision-makers, each with two possible alternatives, provides a simple yet illustrative scenario to study noncooperative interactions. This scenario has been traditionally presented as a payoff table, as defined below.

		Decision-maker 2	
		Decision 1	Decision 2
Decision-maker 1	Decision 1	$u_1(a_{1,1}, a_{2,1}), u_2(a_{1,1}, a_{2,1})$	$u_1(a_{1,1}, a_{2,2}), u_2(a_{1,1}, a_{2,2})$
	Decision 2	$u_1(a_{1,2}, a_{2,1}), u_2(a_{1,2}, a_{2,1})$	$u_1(a_{1,2}, a_{2,2}), u_2(a_{1,2}, a_{2,2})$

This formal description of games is usually motivated by real-world situations. There is a long list of games such as battle of the sexes, prisoners' dilemma, matching pennies, and dictator game. As a didactic example, we will analyze next one of those well-known games representing the relation between two decision-makers that could select aggressive and nonaggressive behaviors in a game called dove and hawk.

Example 6.4 *Dove and hawk game* This is a game that represents a situation where decision-makers can select between two alternative behaviors: being aggressive (hawk), or nonaggressive (dove). The idea is to capture a situation that one being aggressive and the other nonaggressive leads to a stable situation, where one decision-maker "exploits" the other but there is no advantage to change because both being aggressive would lead to a fight and both would end up losing. The socially best situation is when both are nonaggressive, and thus, the decision-makers would coexist in peace. However, being nonaggressive is never a solution for the game-theoretical individuals because changing the behavior would lead to an increase in the individual payoff.

The game can be represented in normal form as follows.

		Decision-maker 2	
		Aggressive	Nonaggressive
Decision-maker 1	Aggressive	$u_1 = x, u_2 = x$	$u_1 = w, u_2 = y$
	Nonaggressive	$u_1 = y, u_2 = w$	$u_1 = z, u_2 = z$

We can analyze the situation based on the relation between payoffs as

- both being aggressive is worse than both being nonaggressive, therefore: $x < z$,
- being aggressive gives advantage in relation to be nonaggressive: $w > y$, and
- there shall be Nash equilibria when one player is aggressive and the other is nonaggressive. In this sense: $w > z$ and $y > x$.

Putting all the above together, we have $x < y < z < w$. We can test this with numbers.

		Decision-maker 2	
		Aggressive	Nonaggressive
Decision-maker 1	Aggressive	$u_1 = 0, u_2 = 0$	$u_1 = 5, u_2 = 1$
	Nonaggressive	$u_1 = 1, u_2 = 5$	$u_1 = 4, u_2 = 4$

By inspection we can see that: We can analyze the situation based on the relation between payoffs as

- Assuming that decision-maker 2 is aggressive, the best alternative for decision-maker 1 is to be nonaggressive.
- Assuming that decision-maker 2 is nonaggressive, the best alternative for decision-maker 1 is to be aggressive.
- Assuming that decision-maker 1 is aggressive, the best alternative for decision-maker 2 is to be nonaggressive.
- Assuming that decision-maker 1 is nonaggressive, the best alternative for decision-maker 2 is to be aggressive.

There are two Nash equilibria, one aggressive and the other nonaggressive. Therefore, this is an example of an anti-coordination game where the situation in which both selecting the same alternative is always unstable.

Because of such an instability, this game is used in evolutionary game theory [16], which tries to explain animal behavior using game theory, in a model introduced in a seminal paper called *The Logic of Animal Conflict* [17]. The idea is to verify the dynamic changes of strategy when the game is played for different rounds, which lead to different sequences of behavior, thus resulting in long-term strategies like retaliation or bullying.

Dove and hawk game also indicates the possibility of sequential games, where one decision-maker is the first mover, and the second reacts to the first mover. This is usually represented by game in *extensive form*, which also helps to analyze the situation where there is the possibility to randomize the decisions through *mixed strategies*, which are defined next.

Definition 6.3 *Games in extensive form, mixed strategy and sequential games* Games in extensive form are represented through a tree topology as depicted in Figure 6.6. There is a possibility to randomize the selected strategy so that decision-maker 1 could select decision 1 with the probability p_1 and decision 2 with the probability $1 - p_1$. Likewise, decision-maker 2 has a mixed

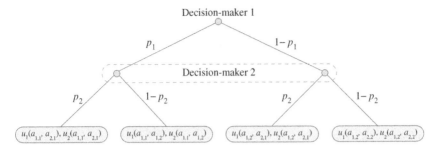

Figure 6.6 Game in the extensive form.

strategy associated with p_2 for decision 1 and $1 - p_2$ for decision 2. Figure 6.6 also shows this. A *mixed strategy Nash equilibrium* refers to a situation where deviations in the individual decision-makers' mixed strategies cannot provide gains in their own *expected* payoff considering that all the other decision-makers are in equilibrium.

A *sequential game* is defined when each decision-maker consecutively selects their own choices considering that the later elements have some information of the earlier ones. In the games with two decision-makers, if decision-maker 1 is the first to decide, decision-maker 2 will have some information about the decision made by the first one. The Nash equilibrium is found through backward induction, where the sequence of payoff-maximizing decisions is found from the leaves to the root of the decision tree.

Example 6.5 *Dove and hawk: extensive form, mixed strategies and sequential game* The dove and hawk game is presented in Figure 6.7 in its extensive form with mixed strategies.

To find the Nash equilibrium, we need to analyze the situations of the two decision-makers. Let us start with decision-maker 2.

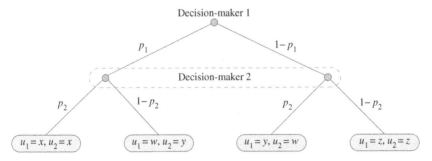

Figure 6.7 Dove and hawk game in the extensive form.

- If it decides to be aggressive, the expected payoff is related to the mixed strategy of decision-maker 1, resulting in $p_1 x + (1 - p_1)w$.
- If it decides to be nonaggressive, the expected payoff is $p_1 y + (1 - p_1)z$.
- To be in equilibrium, both expected payoffs must have the same value, and therefore $p_1 x + (1 - p_1)w = p_1 y + (1 - p_1)z$.

As the game is symmetric and not sequential, we can proceed likewise for decision-maker 1 and then, show that $p_2 x + (1 - p_2)w = p_2 y + (1 - p_2)z$. By solving the equations, we have:

$$p_1 = p_2 = \frac{z - w}{x + z - y - w}, \tag{6.4}$$

where $x < y < z < w$ because this is a dove and hawk game. Note an interesting thing: by being in the Nash equilibrium, the analysis of decision-maker 2 provides the mixed strategy of decision maker 1, and vice versa.

Solving the problem for the situation described in Example 6.4, we have a mixed strategy indicating that each decision-maker randomizes it following a probability 0.5 (i.e. they select to be aggressive with a probability of 50%). However, we can modify the benefits of being aggressive while the other is not aggressive by varying w from 5 to 11, while keeping the other parameters as they were before. The new Nash equilibrium is being aggressive with 87.5% probability, and thus, being nonaggressive with 12.5% probability.

If we assume the game to be sequential where, without loss of generality, decision-maker 1 is the first to decide followed by decision-maker 2, who knows the other's choice. Through backward induction, decision-maker 1 will assess how decision-maker 2 will decide. We know that dove and hawk is an anticoordination game, and thus, decision-maker 2 will always select the alternative that is different from the one previously selected by decision-maker 1. Knowing this, decision-maker 1 will select the alternative that leads to a higher payoff, namely being aggressive. The Nash equilibrium in this situation is decision-maker 1 being aggressive, and decision-maker 2 being nonaggressive. This indicates the advantage that the first mover has.

Game theory as briefly introduced here has an axiomatic assumption that all decision-makers follow only one rule: maximize the individual payoff. Mixed strategies indicate the possibility of randomization of decisions using, for example, a random number generator, a coin, or an urn. For example, in the dove and hawk with 50% probability as its Nash equilibrium, each decision-maker could throw a fair coin having the following decision rule:

- if tail, then select aggressive,
- if head, then select nonaggressive.

Nevertheless, the individual objective is still the maximization of expected pay-offs after repeated games.

Note that it is possible to define many other decision rules, and payoff maximization is not the only option. In the following section, we will present more details of rule-based decision-making.

6.5 Rule-Based Decisions

Every decision-making process is based on rules. For instance, a decision-maker who is willing to select the operating point that will lead to the optimal use of resources has an implicit rule: always select the operating point that will lead to such an optimal outcome. Likewise, the game-theoretic decision-makers are always ruled by payoff maximization. Rule-based decisions are a generalization of the previous, excessively restrictive, rules.

Decision rules can generally be represented by input and output relations. For example, a decision-maker could use the logic gate AND to select an alternative. In concrete terms, a given person will go to the restaurant if his/her three friends will go. After calling and confirming that all three go, the person decides to go. Another option is to use an OR gate, and thus, at least one of the friend needs to go.

Other cases could be proposed. Imitating decisions from other decision-makers or repeating the same decision until some predetermined event happens are two common examples of rules of thumb. Another example is the preferential attachment process described in the previous chapter, where new nodes in a network select with higher chances to be connected to nodes already with a high degree. In the following, we will exemplify a rule-based decision process designed to indicate the potential risks.

Example 6.6 *Automated alarm system* Consider the following scenario: a city with two weather stations capable of measuring temperature and wind speed. Experts indicate that there are a few events that are potentially dangerous: (A) temperature above thirty degrees Celsius, (B) temperature below $-20\,°C$, and (C) wind speed above 70 km/h. Besides, they have also pointed out that the most dangerous situation is when events B and C occur in the same place. The proposed solution is illustrated in Figure 6.8. For example, if events A and C happen at the same time based on the measurements in station 1, the local station will flag a risk level 2. But, if B and C occur at the same time, the local station will indicate a risk level 3, and also the city monitoring station that processes the data from the two stations will indicate a high risk situation in the city. If both stations indicate a risk level 3, then the risk level of the city will be the highest possible.

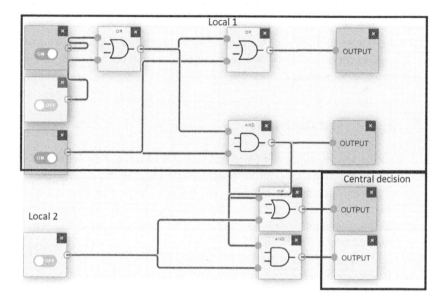

Figure 6.8 Example of the implementation of the alarm system using logic circuits. Generated through https://academo.org/demos/logic-gate-simulator/.

This simple example illustrates well the idea of rule-based decision-making that is usually employed in sensor networks [18]. Besides, decision systems designed to support experts (i.e. expert systems), which are widespread today, relies on rules [19]. Heuristics are another approach that is usually based on predefined or "learned" rules.

We will return to this topic in later chapters. Now, it is important to discuss the limitations of the proposed methodologies for decision-making.

6.6 Limitations

Decision-making processes, as defined here, are always data-mediated processes governed by decision rules, which are flexible and not universal. Decision rules are normative and open for changes, never inviolable as physical laws. This imposes a fundamental limitation because the outcome of decision-making processes can only be predicted if decision rules are known, assumed to be known, estimated, or arbitrarily stated. Even worse, this is only a necessary condition, not sufficient: even if the decision rules are known, the outcome of the decision-making process can only be determined in few special cases.

In this case, the mathematization introduced in Sections 6.3 and 6.4 refers to particular decision-making processes supported by specific assumptions of decision rules, and thus, they are not universal and natural as frequently claimed. This fact is different from saying that optimization and game theory are useless or irrelevant, but only to emphasize that their reach is limited, and the mathematically consistent results must always be critically scrutinized as scientific knowledge. It is important to repeat here: mathematical formalism is neither a necessary nor a sufficient condition to guarantee that a piece of knowledge is scientific.

Besides, decision-making processes depend on structured and interpretable data as inputs. Except for particular decision rules that are independent of the input data (e.g. random selection between alternatives, or always select the same alternative regardless of the input), all decisions are informed. Hence, decision-makers need data with the potential of being information about the possible alternatives of actions to be taken by the agents. Informed decision-making processes need data to decrease uncertainty, and thus, lack of data, unstructured data, misinformation, and disinformation are always potential threats to them.

In this case, each logical link between different decision-makers, or between decision-makers and agents, creates a potential vulnerability for the decision-making process, whose impact depends on its topological characteristics. Centralized decision-making relies on one decision-maker, and thus, the process is vulnerable to the operation of such an element. Decentralized and distributed decision-making processes are dependent on their network topology, but they are usually more resilient and robust against attacks. However, depending on the deliberation method used by decision-makers (which may be governed by different individual decision rules) in those topologies, a selection between alternatives may never happen and the decision-making process may never converge to a choice.

This brings in another limitation: individual decision-makers need to have access to a rule book or doctrine that states how to determine the selection of one (or a few) alternatives among all possibilities. Moreover, in many cases of distributed and decentralized decision-making processes, there should be an additional protocol involving a decision-making process with many decision-makers from the same or different levels. In game theory, this is exemplified by the knowledge of the rules of the game. The problem is even more challenging if we consider decision-making processes of higher order where there is a decision-making process to decide which rule book will be used to decide. This might lead to reflexive or self-referential processes that are undecidable, such as the liar paradox and the halting problem studied by Turing [20].

Those limitations are the key to scientifically understand "autonomous" decision-making processes, which are the basis of cyber-physical systems, thus avoiding design pitfalls and tendencies of universalization, which too often haunt cybernetics as elegantly described in [21]. We should stop at this point, though, and wait for the chapters to come where the theory of cyber-physical system will be presented, and the main limitations of decision-making processes will be revisited.

6.7 Summary

This chapter introduced decision-making processes performed by decision-makers that will select, among possible alternatives, one action or a set of actions that agents will perform. We classified decision-making processes with respect to the topological characteristics of the network composed of decision-makers and agents. We also showed different approaches to analyze and perform decision-making, namely optimization, game theory, and rule-based decisions. Overall, our aim was to show that decision-making processes in cyber-physical systems are (i) mediated using potentially informative data as inputs and selecting action(s) to be performed by agent(s) as output, (ii) governed by decision rules or doctrines that are neither immutable as physical laws nor "natural" as usually claimed, (iii) susceptible to failures, and (iv) not universal.

Despite the immense specialized literature covering fields as diverse as evolutionary biology, psychology, political economy, and applied mathematics, there are few general texts that are worth reading. The paper [5] provides a well-grounded overview of different approaches to human decision-making in contrast to the mainstream orthodoxy [2, 4]. In a similar manner, the book [15] makes a critical assessment of game theory, indicating how its models and results should be employed with the focus on political economy. The paper [22] is an up-to-date overview of economics considering a more realistic characterization of human decision-makers. The seminal paper [17] on explaining the evolution of animal behavior by using game theory provides interesting insights into the fields of theoretical biology and evolutionary game theory. In a technical vein, which is more related to the main focus of this book, the paper [19] can provide an interesting historical perspective of the early ages of expert systems. In their turn, the papers [7–9] discuss issues on decentralized and distributed decision-making systems in the late 2010s and early 2020s.

Exercises

6.1 **Wireless networks and decision-making** Compare the scenario presented in Example 6.3 with another considering that each transmitter acts selfishly in other to maximize its own link throughput $T = p \exp(-kp\lambda)$. The optimization problem can be formulated as:

$$
\begin{aligned}
\max_{p} \quad & p \exp(-kp\lambda) \\
\text{s.t.} \quad & p \leq 0 \\
& p \geq 1
\end{aligned}
\tag{6.5}
$$

(a) What is the optimal solution p^* and $T^* = p^* \exp(-kp^*\lambda)$?

(b) Plot the spatial throughput $S = \lambda T^*$ obtained when all the transmitters are selfish for $\lambda = 0.25$, $\lambda = 2.5$ and $\lambda = 25$ for $k = 1$.

(c) Discuss the results.

6.2 Prisoners' dilemma Prisoners' dilemma is a well-known game describing a situation where independent decision-makers, which cannot communicate with each other, are stuck in a Nash equilibrium that is noncooperative, although it seems to be their best interest to be cooperative. The problem is described as follows. Two members A and B of a criminal gang are arrested and imprisoned. Each prisoner is in solitary confinement with no means of communicating with the other. The prosecutors lack sufficient evidence to convict the pair on the principal charge, but they have enough to convict both on a lesser charge. Simultaneously, the prosecutors offer each prisoner a bargain. Each prisoner is given the opportunity either to betray the other by testifying that the other committed the crime, or to cooperate with the other by remaining silent. The possible outcomes are:

- If A and B each betray the other (not cooperating with each other), each of them serves z years in prison (payoff of $-z$).
- If A betrays B (not cooperating with B) but B remains silent (cooperating with A), A will serve y years in prison (payoff $-y$) and B will serve w years (payoff of $-w$).
- If B betrays A (not cooperating with A) but A remains silent (cooperating with B), B will serve y years in prison (payoff $-y$) and A will serve w years (payoff of $-w$).
- If A and B both remain silent, both of them will serve x years in prison (payoff of $-x$).

Member A

		Cooperate	Betray
Member B	Cooperate	$u_1 = -x, u_2 = -x$	$u_1 = -w, u_2 = -y$
	Betray	$u_1 = -y, u_2 = -w$	$u_1 = -z, u_2 = -z$

(a) What is the relation between the payoff values $x \geq 0$, $y \geq 0$, $w \geq 0$ and $z \geq 0$ so that the game can be classified as a Prisoner's Dilemma? Note: Follow the approach used in Example 6.4 and note that the payoffs in the table are negative but the values of x, y, w, z are positive.

(b) Would this game be a dilemma if the two members were trained to not betray each other, and thus, they have a background knowledge about how the other would decide? Justify and analyze this situation as if you were one of the captured persons.

(c) Study prisoners' dilemma as a sequential game following Example 6.5.

(d) Repeated prisoners' dilemma is usually used to study the evolution of cooperation in competitive games. In the abstract of the seminal paper [23] that introduces this idea, we read: *Cooperation in organisms, whether bacteria or primates, has been a difficulty for evolutionary theory since Darwin. On the assumption that interactions between pairs of individuals occur on a probabilistic basis, a model is developed based on the concept of an evolutionarily stable strategy in the context of the Prisoner's Dilemma game. Deductions from the model, and the results of a computer tournament show how cooperation based on reciprocity can get started in an asocial world, can thrive while interacting with a wide range of other strategies, and can resist invasion once fully established. Potential applications include specific aspects of territoriality, mating, and disease.* The task here is to critically assess the statements made by this text based on the discussions introduced in this chapter.

6.3 **Alarm system** Consider the scenario presented in Example 6.6. We assume that events A, B, and C are associated with a probability p_A, p_B, and p_C. The following table defines the risk levels at the local stations.

Event A	Event B	Event C	Risk level
True	False	True	Level 2
False	True	True	Level 3

For the city level, the risk levels are given next.

Local 1	Local 2	Risk level
Level 3	Level -	Level 2
Level -	Level 3	Level 2
Level 3	Level 3	Level 3

All the other situations are classified as risk level 1.

(a) Compute the probability mass function of the risk levels at the local stations. Note: events A and B are mutually exclusive (i.e. they cannot be true at the same time).

(b) Compute the probability mass function of the risk level at the central station considering that the two local stations are independent of each other.

(c) Critically assess the results presented in (b) based on the assumption of independence.

(d) Determine the level that each decision-making process has following Proposition 6.1.

References

1 Merriam-Webster Dictionary. Decision; 2021. Last accessed 1 March 2021. https://www.merriam-webster.com/dictionary/decision.

2 Steele K, Stefánsson HO. Decision Theory. In: Zalta EN, editor. The Stanford Encyclopedia of Philosophy. winter 2020 ed. Metaphysics Research Lab, Stanford University; 2020.

3 Mirowski P. Against Mechanism: Protecting Economics from Science. Rowman & Littlefield Publishers; 1992.

4 Wolff RD, Resnick SA. Contending Economic Theories: Neoclassical, Keynesian, and Marxian. MIT Press; 2012.

5 Gigerenzer G. How to explain behavior? Topics in Cognitive Science. 2020;12(4):1363–1381.

6 Srnicek N. Platform Capitalism. John Wiley & Sons; 2017.

7 Baldwin J. In digital we trust: Bitcoin discourse, digital currencies, and decentralized network fetishism. Palgrave Communications. 2018;4(1):1–10.

8 Pournaras E, Pilgerstorfer P, Asikis T. Decentralized collective learning for self-managed sharing economies. ACM Transactions on Autonomous and Adaptive Systems (TAAS). 2018;13(2):1–33.

9 Nguyen MN, Pandey SR, Thar K, Tran NH, Chen M, Bradley WS, et al. Distributed and democratized learning: philosophy and research challenges. IEEE Computational Intelligence Magazine. 2021;16(1):49–62.

10 Boyd S, Vandenberghe L. Convex Optimization. Cambridge University Press; 2004. Available at https://web.stanford.edu/~boyd/cvxbook/bv_cvxbook.pdf.

11 Popovski P. Wireless Connectivity: An Intuitive and Fundamental Guide. John Wiley & Sons; 2020.

12 Nardelli PHJ, Kountouris M, Cardieri P, Latva-Aho M. Throughput optimization in wireless networks under stability and packet loss constraints. IEEE Transactions on Mobile Computing. 2013;13(8):1883–1895.

13 Leyton-Brown K, Shoham Y. Essentials of game theory: a concise multidisciplinary introduction. Synthesis Lectures on Artificial Intelligence and Machine Learning. 2008;2(1):1–88.

14 Syll LP. Why game theory never will be anything but a footnote in the history of social science. Real-World Economics Review. 2018;83:45–64.

15 Hargreaves-Heap S, Varoufakis Y. Game Theory: A Critical Introduction. Routledge; 2004.

16 Vega-Redondo F. Evolution, Games, and Economic Behaviour. Oxford University Press; 1996.

17 Smith JM, Price GR. The logic of animal conflict. Nature. 1973; 246 (5427): 15–18.

18 Nardelli PHJ, Ramezanipour I, Alves H, de Lima CH, Latva-Aho M. Average error probability in wireless sensor networks with imperfect sensing and communication for different decision rules. IEEE Sensors Journal. 2016;16(10):3948–3957.

19 Hayes-Roth F. Rule-based systems. Communications of the ACM. 1985;28(9):921–932.

20 Turing AM. On computable numbers, with an application to the Entscheidungsproblem. Proceedings of the London Mathematical Society. 1937;2(1):230–265.

21 Gerovitch S. From Newspeak to Cyberspeak: A History of Soviet Cybernetics. MIT Press; 2004.

22 Arthur WB. Foundations of complexity economics. Nature Reviews Physics. 2021;3:136–145.

23 Axelrod R, Hamilton WD. The evolution of cooperation. Science. 1981;211(4489):1390–1396.

Part II

7

The Three Layers of Cyber-Physical Systems

This chapter introduces the three layers that constitute every cyber-physical system (CPS). Our aim is to articulate the concepts that were presented in the first part of this book in order to construct a scientific theory of CPSs that are systematically characterized by physical, data, and decision layers, as well as the necessary processes that define cross-layer relations. One important remark before we start: there are several layered architectures to analyze and develop specific CPSs, such as industrial production systems [1, 2] and power grids [3], but our proposal is very different. Those existing layered models focus on practical questions of particular cases, offering flexibility and a unified lexicon to assess and deploy such CPSs. In this book, our objective is to characterize what constitutes CPSs as such, not restricting our study to specific cases. In fact, different CPSs will be analyzed in Part III as realizations of the general theory proposed here. The approach taken here extends the ideas introduced in [4–7].

7.1 Introduction

In the first chapter, we provided a broad definition of CPSs, which was put forth by the standardization organization NIST. This chapter will finally state our proposed definition of CPSs. The idea is to characterize a typical CPS following the concept of "system" as defined in Chapter 2, indicating that, to be a CPS, it must be composed of a physical system from where data are collected to be processed and then used in a decision-making process that will determine actions that have the potential to modify the physical system. Therefrom, we can define the physical, data, and decision layers of CPSs, and also measuring/sensing, informing, and acting cross-layer processes. In what follows, we will formalize this definition.

Cyber-physical Systems: Theory, Methodology, and Applications, First Edition. Pedro H. J. Nardelli.
© 2022 The Institute of Electrical and Electronics Engineers, Inc. Published 2022 by John Wiley & Sons, Inc.

Definition 7.1 *Cyber-physical system* Particular systems whose peculiar function only depends on physical relations to operate are denominated physical systems. In contrast, particular systems that are necessarily constituted by physical and logical relations to perform their peculiar function are denominated CPSs. Hence, CPSs necessarily operate mediated by data processes that determine logical relations directly or indirectly employed in decision-making processes associated with actions to maintain their proper functioning. This constitutive mediation by data processes is what defines the cyber domain of CPSs, and thus, the concept of CPSs as such.

This definition states the reasons why CPSs ought to be studied as a specific object that cannot be reduced to other existing theories, such as system dynamics, control theory, information theory, computer sciences, or game theory. Following the concepts introduced in the previous chapters, a full account of a given CPS needs to consider how data are acquired from physical processes, which are related to specific observable attributes, to be then transmitted, stored, and manipulated using information and communication technologies (ICTs). These processed data should then become informative about the operation of the system under consideration allowing for informed decision-making processes about interventionist actions to guarantee its functioning. Following the approach introduced in [4] and further developed in [7], the simplest theoretical model of every CPS must consider three constitutive layers and three cross-layer processes. The following proposition defines them.

Proposition 7.1 *Three layers of all CPSs* All CPSs are constituted by at least three layers and three cross-layer processes as follows.

- **Physical layer** refers to material systems that are related to physical realities and level 0 processes. It is constituted by direct relations whose dynamics are determined and constrained by physical laws.
- **Measuring or sensing** refers to the cross-layer process that maps physical-layer phenomena into measured and/or sensed attributes. This process determines the relation between the physical and data layers.
- **Data layer** refers to symbolic processes of levels greater than 0 and grounded in the physical layer. It is constituted by data, data processes, and logical relations, which are potentially unbounded but always subjected to energy and information limits. This layer is associated with the *cyber* domain of CPSs.
- **Informing** refers to the cross-layer process of communicating potentially informative data to decision-makers and agents. Note that informing processes can happen among decision-makers and between decision-makers and agents.

- **Decision layer** refers to decision-making processes about actions concerning possible interventions in all three layers and cross-layer processes.
- **Acting** refers to actions taken by agents following the informative data generated through decision-making processes.

This proposition presents how every CPS is organized within its boundaries in the sense that was defined in Chapter 2. It is important to note that CPSs have relations with the external world, including other systems and the environment in general. These external conditions can be associated with any layer. Remember that Proposition 7.1 refers to CPSs in general, but the aforementioned relations ought to be defined for their particular realizations.

In addition, CPSs involve more components than previously described in Chapter 2, where we defined the structural, operating, and flow components. For the CPS, we need to extend these to explicitly incorporate *measuring components*, *computing components*, and *communication components* that are necessary to construct CPSs. To illustrate how to analyze a particular CPS using Proposition 7.1 as well as the types of components that we just introduced, we will revisit Example 3 from Chapter 2 as follows.

Example 7.1 *Wind turbine as a CPS* Consider the conceptual model of a new wind turbine to be deployed in apartment buildings. The development of this system incorporates new ICTs to improve its operation. In this case, the physical system has some of its operational attributes measured by sensors, which will inform actuators and a control unit about the behavior of the system. This conceptual wind turbine can be demarcated as follows.

PS (a) Structural components: metal tower, new type of blades, connection to the grid; (b) operating components: power electronic devices and an electric generator; (c) flow components: wind and electricity; (d) measuring components: wind speed sensor, vibration sensor, accelerometers, position sensors, and electric sensors; (e) computing components: data acquisition module, data storage, and data processing unit; (f) communication components: cables connecting different components, and transmission and reception modules.

PF Convert kinetic energy from wind (the main input of the system) into electric energy (the main output) to be used in the apartment building.

C1 It is physically possible to convert the kinetic energy from wind into electric energy that will be used to supply the electricity demand in the building.

C2 The electricity generated has to be synchronized with the grid in case of alternating current (AC) with the same frequency by using power electronic devices as actuators based on the electrical measurements acquired, stored, manipulated, and communicated; proper maintenance of the components indicated by a predictive algorithm that fuse all the data acquired is ensured; there is wind with enough kinetic energy to allow for the power conversion that can be evaluated by the wind speed measurements; there are protective devices against risk situations such as strong winds and overcurrent that can be monitored based on the acquired data.

C3 Battery storage in the building, management of electricity demand, investment programs to support renewable energy, elections, subsidies, willingness to use wind turbines, regulations and laws that allow distributed energy resources, raw material and industries to produce such a new turbine.

Following Proposition 7.1, we can characterize this wind turbine as a CPS by defining its three layers and cross-layer processes.

- **Physical layer:** The wind turbine components that are combined to perform its peculiar function.
- **Measuring or sensing:** The processes performed by the measuring components and sensors used to map specific properties in the physical layer to the data layer.
- **Data layer:** It refers to the cyber (symbolic) domain composed of digital or analog measured data that can be processed and manipulated by computing components in order to convert them into information about specific processes and subsystems related to the operation of the wind turbine.
- **Informing:** The processes of communicating potentially informative data to different elements of the cyber domain including decision-makers and agents; this would involve the transmission of machine interpretable data following specified communication protocols, dashboards as a human–computer interface, and machine learning predictive methods to be used as inputs in fault detection algorithms.
- **Decision layer:** It involves decision-makers, being them humans or software-defined elements, that have the role of determining the interventions to be taken based on potentially informative data to guarantee the proper operation of the wind turbine; for example, a machine learning algorithm can predict that there will be a potential fault in the rotor during the next five hours, or that the energy to be produced in the next ten minutes will be above the expected demand so that it should be stored in the battery system.
- **Acting:** Agents, which can be either humans or machines, informed of the decisions should act to modify or intervene in their respective domain of action. For example, an actuator can modify a switch based on the input data to store the energy generated by the wind turbine, or a person can call the maintenance team after receiving an alarm message about an expected fault.

Roughly speaking, the conceptual turbine that we just presented is a CPS because the cyber domain constituted by informative data is necessary to its successful physical operation, namely converting wind into electricity. The cyber domain is also supported by physical devices – the measuring, computing, and communication components. However, the cyber domain itself is constituted by meaningful data and logical relations, both related to semantics. The regulation of the physical process of energy conversion is then mediated by potentially informative data that are the basis of operation decisions and the respective interventions. In the following sections, we will present in more details the three layers of CPSs and the respective cross-layer processes.

7.2 Physical Layer, Measuring, and Sensing Processes

A presupposition of any engineering conceptual system is that it has the potential to be materially deployed. The wind turbine exemplified in the previous section is designed aiming at its physical construction in the future. Cars, highways, bikes, microwave ovens, brain–computer interfaces, televisions, wireless network base stations, mobile phones, and power grids have a material existence, and thus, their behavior at this layer can be evaluated by physical laws. The level of details is dependent on the specific application. For an incandescent lamp, the simple characterization of electric circuits based on Ohm's law is enough. If the system under consideration is an electronic circuit with transistors, the design would require a more detailed analysis of the electric field in different materials. Currently, emerging topics of quantum computing [8] and biological implementation of logical gates [9] consider, in their own way, different levels of characterization based on fundamental physical laws.

The physical layer of CPSs refers to those aspects that are governed by direct physical relations, which are different from the logical relations that are mediated by data. Figure 7.1 illustrates a simple representation of a direct current electric circuit whose full description is derived from the well-known Ohm's law: $V = RI$ (i.e. the voltage between two given points is equal to the current that passes through a conductor between these two points times its electric resistance); the dissipated power by the resistor is computed as $P = VI$. As indicated by the figure, there is a switch before each resistor, which could be on (closed) or off (opened). In this case, the circuit could have different physical topologies that lead to different system behavior, which will be presented next. For the calculations, the values of the resistances are the same $R_1 = R_2 = R_3 = R$, and the source produces a constant voltage V. We are interested in the power dissipated by each resistor P_1, P_2, and P_3, as well as by the whole circuit $P = P_1 + P_2 + P_3$.

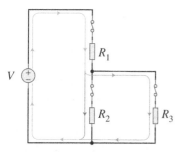

Figure 7.1 Electric circuit with direct current: a constant voltage source V and three resistors R_1, R_2, and R_3.

Case 1: The switch of R_1 is off, or the switches of R_2 and R_3 are off. In both cases, the equivalent resistance is $R_{eq} = \infty$ so that $I = 0$. Without current, no power is dissipated ("consumed") by the resistors. Then, $P_1 = P2 = P_3 = 0$.

Case 2: The switch of R_1 is on, and only one of the other switches from either R_2 or R_3 is off. In this scenario, the resistors will be connected in series and the equivalent resistance is $R_{eq} = 2R$. This will result in $I = V/(2R)$. Since the resistors have the same resistance, the power dissipated by each resistor is $P_1 = V^2/(4R)$, and $P_2 = V^2/(4R)$ and $P_3 = 0$, or $P_2 = 0$ and $P_3 = V^2/(4R)$. The total dissipated power is $P = V^2/(2R)$.

Case 3: All the three switches are on. In this scenario, resistors R_2 and R_3 are connected in parallel, resulting in an equivalent of $R/2$, which is connected in series with R_1. This will lead to an equivalent $R_{eq} = 3R/2$. This will result in a current in the equivalent circuit $I_{eq} = 2V/(3R)$ and a total power dissipated $P = 2V^2/(3R)$. The current passing through R_1 is I_{eq}, and thus, $P_1 = RI_{eq}^2 = 4V^2/(9R)$. Then, $P_2 + P_3 = P - P_1$. As R_2 and R_3 have the same value, $P_2 = P_3 = V^2/(9R)$.

These equations unambiguously determine the behavior of the electric circuit. One important aspect here is to explicitly mention that these relations presume the electric phenomena under investigation to be measurable. Consider that the voltage in the source is the input of this circuit, and the output of the system is the dissipated power, which is measured in the three resistors. Figure 7.2 numerically exemplifies cases 2 and 3 considering an arbitrary setting.

In this case, we consider a perfect measuring process based on a conceptual example. In real-world devices there are many problems in measuring (and sensing), such as poor calibration, interference, and noise. Regardless of those challenges, measuring and sensing processes are necessary to map physical processes into the data. This is what is usually called *data acquisition*.

In CPS, data acquisition is almost always digital, not analog. In this case, the data acquisition is always quantized in time (not continuous), being it periodic (i.e. acquire new data sample every 10 seconds), event-driven (i.e. acquire new data sample when a predetermined event happens, e.g. the dissipated power is above a threshold value), or a hybrid between both. The acquired data are hence stored, transmitted, and processed in bits, which are constitutive of the data layer, which will be discussed next.

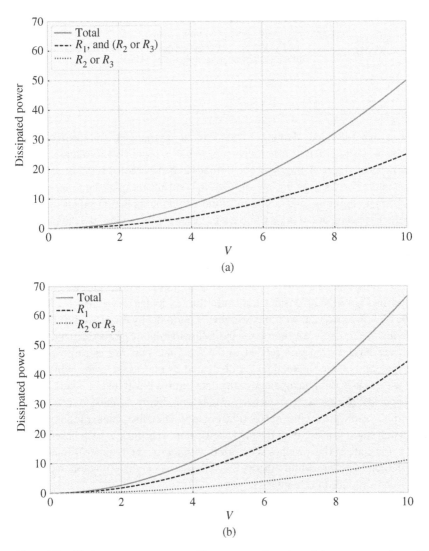

Figure 7.2 Dissipated power measured in watts as a function of the constant voltage V in volts for the circuit presented in Figure 7.1 for $R_1 = R_2 = R_3 = 1$ Ω. (a). Case 2. (b) Case 3.

7.3 Data Layer and Informing Processes

After sensing and measuring processes, the physical layer can be mapped to another, symbolic, domain called data layer. The data layer constitutes a cyber reality where new forms of relations – the logical relations – are established

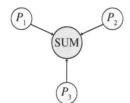

Figure 7.3 Data layer with four cyber elements: three representing the dissipated power P_1, P_2, and P_3 by resistors R_1, R_2, and R_3 from Figure 7.1 and one representing the sum of values obtained from the other three.

between cyber-defined elements. Data can be exchanged and processed in different ways by different cyber elements. What is important to remark is that the data layer refers to the symbolic domain and data processes that are level 1 or greater. The focus is on logical relations and meaningful data, not on the measuring, computing, and communication components that are their material support. Therefore, and it is very important to keep this in mind, the data layer is at the last instance about semantics with respect to the cyber elements and their logical relations whose ultimate objective is to produce data that are potentially informative in decision-making processes at the decision layer.

Let us illustrate this idea by considering a system whose physical layer is defined in Figure 7.1. Consider that only the dissipated power P_1, P_2, and P_3 can be directly observed, with measurements periodically acquired in time with the index $k = 0, 1, \cdots$. We then have the following data $P_1[k]$, $P_2[k]$, and $P_3[k]$, which are level 1 processes related to the kth time step. We can also process these data and obtain the total power dissipated by the circuit as the sum $P[k] = P_1[k] + P_2[k] + P_3[k]$. The topology of the data layer is illustrated in Figure 7.3.

We consider the scenario where each switch is randomly assigned to state a *on* or *off* with the same probability of 50% (like tossing a fair coin). Then, the changes in the switch states is faster than the data acquisition, and thus, at each observation time k, the state of the switches is on or off with a probability of 50%. To analyze this problem with probability theory, we will first generalize it by following the approach introduced in Chapter 3.

- System T: Electric circuit depicted in Figure 7.1.
- Protocol ρ: At each time index k, $P_1[k]$, $P_2[k]$, and $P_3[k]$ is observed and recorded.
- Attributes $a \in \mathcal{A}$ are the power dissipated by the resistors so that $\mathcal{A} = \mathbb{R}$ (negative power values mean that power is consumed).
- Experiment Ξ: Perform $P[k] = P_1[k] + P_2[k] + P_3[k]$.
- The random variable for the kth observation is: $X(k) = P[k]$ with the sample space $S_X = \mathbb{R}$.
- The stochastic process is defined as $\{X(0), X(1), \cdots \}$, which is memoryless.

To numerically exemplify this situation, we assume that $V = 1$ V, and $R_1 = R_2 = R_3 = 1 \, \Omega$. In this situation, the sample space can be reduced to three outcomes, which are the three cases described in the previous section. Hence, the sample space is $S_X = \{0, 1/2, 2/3\}$ related to case 1, 2, and 3, respectively. If nothing is known about this probability distribution of $X(k)$, we calculate its maximum information entropy as introduced in Chapter 4: $H_{\max,X} = \log_2 3 = 1.58$ bits. However, we know how to map the probabilities of the three events of the sample space by studying the combination of the switch states that result in cases 1, 2, and 3, as follows.

- Case 1: R_1 off, or R_1 on and R_2 and R_3 off, then $P(X(k) = 0) = p_1 = 0.5 + 0.5^3 = 0.625$.
- Case 2: R_1 on, and only one of R_2 or R_3 off, then $P(X(k) = 1/2) = p_2 = (0.5)^3 + (0.5)^3 = 0.25$.
- Case 3: All on, then $P(X(k) = 2/3) = p_3 = (0.5)^3 = 0.125$.

Figure 7.4 illustrates a realization of the stochastic process with $k = 0, \cdots, 30$.

From the probability mass function just determined, we can calculate the entropy of the random variable X as: $H(X) = -p_1 \log_2 p_1 - p_2 \log_2 p_2 - p_3 \log_2 p_3 = 1.3$ bits. In this case, $H_{\max,X} - H(X) = 0.28$ bit is the (mathematical) information contained in the description of the observation protocol and the preliminary knowledge of the physical system itself.

This quantification is strictly related to the question of the value that $P[k]$ will assume considering each observation time is an i.i.d. realization of the random variable X that is defined by cases 1, 2, and 3. The uncertainty of $P[k]$ is different from the uncertainty of the particular combination of events (i.e. if the switches are

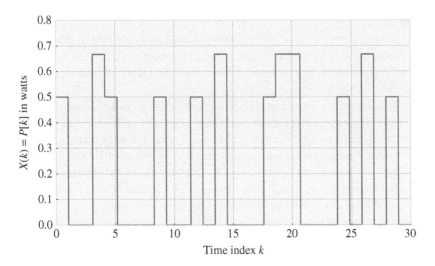

Figure 7.4 Numerical example of a realization of the stochastic processes $P[k]$ (measured in watts) with the time index k that is the outcome of the experiment Ξ related to the physical system T.

on or off) that resulted in such a value. Considering that $P[k]$ is the only observable attribute in the physical system T, there is an unresolvable uncertainty about the state of the system. If $P[k] = 0.625$ W, then we are sure that all switches are on. However, when $P[k] = 0.5$ W, it is impossible to know without additional data if the physical event that resulted in this outcome is R_2 on and R_3 off, or R_2 off and R_3 on. The analysis of this case will be the focus of Exercise 7.7.

There are important remarks here. First, the measured data have potential to be informative in a semantic sense and, in this specific case, they can be evaluated using the mathematical information theory. Second, if the only observable attribute is a level 2 process, there might be an increase in the uncertainty related to the actual state of the physical system that led to such an observable outcome. Third, the specific uncertainty to be resolved is a semantic characterization that should be strictly related to the function of the acquired data in the CPS under consideration. Fourth, other aspects could have been added, such as noisy measurements and communication errors that would increase the uncertainty related to the data processes. In summary, the semantic-functional determination of the acquired and processed data from a given system is what defines what are the possible sources of uncertainty, and thus, which data can be informative to different elements of the CPS.

Another important characterization of the data layer refers to its informing structure, as illustrated in Figure 7.3. Despite the usefulness of the visual and mathematical characterization of networks, we would like to introduce here another form of algebraic formalism to represent the data layer and its constitutive informing processes. The notation is inspired by *the algebra of reflexive processes* introduced by Vladimir Lefebvre in the late 1960s [10] as part of the cybernetics trend of the Soviet sciences [11]. The conceptualization of a reflexive process is given in the first page of Lefebvre's book *Conflicting Structures* [10]:

> What is a reflexive system? Let us use the following analogy. Imagine a room full of crooked mirrors placed at various angles to each other, as typical in amusement parks. If a pencil falls from a table, it will be reflected by the mirrors in many whimsical ways; then the reflections will be reflected with numerous distortions, ad infinitum. An avalanche of distorted images will flash around the room. A reflexive system is a system of mirrors reflecting each other over and over again. Each mirror is analogous to a person with a particular position relative to the world. The entire complicated stream of mirrors' mutual reflections is an analogue of the reflexive process.
>
> The example with the pencil illustrates the difference between physical processes and social-psychological ones. A pencil falling is a physical process. If, however, we are interested not only in this fall but in the entire stream of multiple mirror images, we are dealing with a social-psychological event.

We neither share his philosophical position nor the mathematization proposed in the book. However, the proposed algebraic representation of reflexive systems is a useful tool to characterize informing processes that exist in the data layer of CPSs. The notation is simple, but we need to proceed with care. Consider a physical system T and two elements A and B. If A and B can directly measure or sense T, then the images Ta and Tb are part of the reflexive system; Ta can be read as the image that A has of the physical system T (the capitalization of the fonts is important here). The reflexive system is then $T + Ta + Tb = T(1 + a + b)$, indicating that the images of the measuring elements are part of the system, regardless of its quality and trustfulness. The element $(1 + a + b)$ is called *structure of awareness* of the reflexive system. If we consider a slightly different situation where element B also receives data related to the measurement of A, then $(Ta)b = Tab$ is also part of the reflexive system, which is defined by a new structure of awareness $(1 + a + b + ab)$. Note that in this notation $ab \neq ba$, and that the sum and multiplication operations are defined in a different way in the proposed formalism; for us, it is not important how to operate with the proposed algebraic formalism, but rather to understand the representation it affords.

Although the physical system T is the same in both cases, the two reflexive systems are different, and thus, might result in observable differences in their dynamics, as to be discussed later. What is remarkable is that this simple notation can unambiguously differentiate systems that are constituted by physical and symbolic layers. From the structure of awareness, we can directly infer the level of the data processes by counting the number of reflections needed to access the physical system present in the structure of awareness. In this simple example, there is one level 0 process represented by the physical system T itself, two level 1 processes represented by Ta and Tb, and one level 2 process represented by Tab. It is important to reinforce that the structure of awareness cannot tell whether the reflection is reliable, but only about its existence as part of the reflexive system. This aspect is important to characterize many times hidden differences in logical relations that constitute CPSs and also to assess potential vulnerabilities introduced by the cyber reality.

Let us now study the physical system T and the data layer depicted in Figure 7.3. The reflexive system can be represented based on the level of the process as follows.

- Level 0: The physical system T.
- Level 1: The measurements $Ta + Tb + Tc = T(a + b + c)$ taken by the elements A, B, and C (respectively related to P_1, P_2, and P_3).
- Level 2: The data process $Tad + Tbd + Tcd = T(a + b + c)d$ carried out by the element D, whose function is to sum the measurements of A, B, and C.
- Reflexive system: $T(1 + a + b + c + (a + b + c)d)$.
- Structure of awareness (SAw): $1 + (a + b + c)(1 + d)$.

It is interesting to see that the numerical example presented in this section refers to the uncertainty related to the $T(a + b + c)d$ process itself, while Exercise 7.7 will deal with uncertainties of the reflexive system. However, a complete characterization of CPSs also requires explicit characterization of the decision-making processes and their associated actions, which is the focus of the next section.

7.4 Decision Layer and Acting Processes

After being informed, decision-makers will conduct their respective decision-making processes, which depend on the structure of awareness of the reflexive system, to define the actions to be taken by agents. The different possibilities of decision-making processes are presented in Chapter 6. Decisions concern not only direct physical actions (level 0 processes), such as opening a switch, increasing the voltage, or stopping a production line, but also symbolic actions, such as changing of the decision rules or modification of logical links. Nevertheless, as stated in Chapter 6, all the symbolic actions, directly or indirectly, refer to physical actions depending on their specific process level.

The example of the electric circuit from the two previous sections can be used here to illustrate how the structure of awareness enables different decision-making processes. Hence, the physical layer is the circuit defined in Figure 7.1 and the data layer is the network defined in Figure 7.3 leading to a structure of awareness $SAw_1 = 1 + (a + b + c)(1 + d)$. We consider a centralized decision-making carried out by the element D, which is also the agent that can open and close the switches of the circuit, which is the only possible action that is allowed. If the optimization problem is to maximize the dissipated power $P[k]$, the optimal solution considering that the voltage of the source is fixed is to close all the three switches (i.e. the switches of R_1, R_2, and R_3 are on, which is case 3). After the optimal solution is found, D's decision is defined and it can then act accordingly. Similar to the structure of awareness, we can propose a *structure of action* that can be used to represent the agents of the system. In this example, the active system is dT and the structure of action $SAc_1 = d$ meaning that D is capable of modifying the physical system T.

Actually, the case presented in Section 7.3 where the switches are randomly associated with a state on or off has a different structure of action where nodes A, B, and C are both decision-makers and agents, whose decisions are independent of the state of the system and of each other's decisions. Hence, we have $SAc_2 = a + b + c$ where the individual decision-making process is based on rules related to the random assignments of the on and off states for each switch. In this case, node D is neither an agent nor a decision-maker, only a data processing element.

These two cases are two extreme situations of *reflexive–active systems*, one purely deterministic with centralized decision-making and the other purely random decisions by independent decision-makers. We can also analyze a more intricate scenario where the elements A, B, and C create images of each other to guide their individual decisions. In this case, $SAw_2 = 1 + (a + b + c)(1 + d) + (b + c)a + (a + c)b + (a + b)c$ considering $SAc_2 = a + b + c$. If images that the elements A, B, and C create of each other states that the individual aim is to maximize the individual dissipated power assuming that they know the possible outcomes of the system and that they cannot communicate their decisions, this problem can be studied through game theory. If Element A is off, then the payoff is $(0,0,0)$ regardless of B and C actions. If Element A is on, then we have the following payoff table:

		Element C	
		off	on
Element B	off	$(0,0,0)$	$(1/4, 0, 1/4)$
	on	$(1/4, 1/4, 0)$	$(4/9, 1/9, 1/9)$

It is easy to see that the best response and the pure strategy Nash equilibrium is that all the three elements are on; this is also the global optimum.

Consider a slightly different situation where B and C are now dependent by jointly deciding their actions. One can see that by coordinating their on and off actions (whenever one is on, the other is off), they can improve their average payoffs. The coordination will lead to an average individual payoff of $(1/4)/2 = 1/8$, which is greater than $1/9$. This solution is below the global optimum in terms of total dissipated power and significantly decreases the payoff of A, but improves the average payoffs of B and C and also the fairness of the system. It is worth saying that those two last cases have the same structures of awareness and action, but they are constructed upon different assumptions of how the elements B and C form the images of each other (the first one based on game-theoretic assumptions, the second based on explicit cooperation).

Reflexive–active systems, which have internal structures of awareness and action, allow for self-development where agents governed by decision-makers intervene in the physical system based on informative data used in decision-making processes. The following section will formalize self-developing reflexive–active systems that are the purified version of CPSs.

7.5 Self-developing Reflexive–Active System and Cyber-Physical Systems

The idea of self-developing reflexive–active environment was introduced as a broader foundational concept of the third-order cybernetics [12]. As already indicated in Chapter 1, this book has not been written based on the same philosophical background, but the characterization of systems as self-developing and reflexive–active is appealing and useful for our purposes, and thus, we shall retain it. Our aim here is to formalize it in order to characterize self-developing reflexive–active systems as a purified version of CPSs.

Proposition 7.2 *Self-developing reflexive–active systems* A given self-developing reflexive–active system Φ (whose demarcation follows the procedure introduced in Chapter 2 and illustrated in Example 7.1) is constituted by three layers and cross-layer processes introduced in Proposition 7.1. It is then possible to formalize Φ by determining:

(1) direct causal relations that are usually stated as mathematical relations of input variable(s) and (observable and nonobservable) attribute(s) of the system;
(2) its reflexive elements (specific data processors and decision-makers with their decision rules), and its respective structure of awareness (i.e. logical relations and communication network) and specific data processes;
(3) its active elements (agents), and its respective structure of action and different types of actions.

These three items are necessary and sufficient to formally characterize Φ as a self-developing reflexive–active system constituted by physical, data, and decision layers.

Proposition 7.3 *Relation between self-developing reflexive–active systems and CPSs* Self-developing reflexive–active system is a theoretical, purified object that can be used to conceptualize CPS with its three constitutive layers. Hence, all CPSs can be categorized as self-developing reflexive–active systems. However, the latter is a broader concept than the former, and thus, there exist self-developing reflexive–active systems that are not CPSs.

After the theoretical journey presented in the previous chapters, it is possible to say that the fundamentals of the theory of CPSs are now stated. In the following, we will exemplify how to design a CPS based on the electric circuit studied in the previous sections.

Example 7.2 *Self-developing electric circuit* Consider the electric circuit presented in Figure 7.1. Then, we can characterize the CPS Φ as follows.

(1) **Input:** $V[k]$ in volts, and **output:** $P[k] = V^2/R_{eq}[k]$ in watts where the equivalent resistor $R_{eq}[k]$ in ohms depends on the states (on or off) of the switches A, B, and C controlling the connections of resistors R_1, R_2, and R_3, respectively;

k refers to the time index where the periodic measurement of $P[k]$ is collected from node D. The potential outcomes are indicated in cases 1, 2, and 3. We set $V = 1$ V and $R_1 = R_2 = R_3 = 1$ Ω for numerical evaluation.

(2) **SAw:** $1 + (a + b + c)(1 + d) + bc + cb$. This requires a communication between elements B and C, and from A, B, and C to D. The communication links are error-free serving (i) B and C to periodically exchange, with the same time index k, the data about their individual state $s_B[k]$ and $s_C[k]$, which is either on or off, and (ii) A, B and C to send their individual dissipated power $P_1[k]$, $P_2[k]$ and $P_3[k]$ to D, which calculate $P[k] = P_1[k] + P_2[k] + P_3[k]$.

(3) **SAc:** $b + c$. Elements B and C make autonomous decisions based on the received information; they are also agents that can control their respective switches. To compare the outcomes, we consider three scenarios: (i) $s_B[k + 1] = s_C[k]$ and $s_C[k + 1] = s_B[k]$, (ii) $s_B[k + 1] = $ not $s_C[k]$, and $s_C[k + 1] = s_B[k]$, or (iii) $s_B[k + 1] = $ not $s_C[k]$, and $s_C[k + 1] = $ not $s_B[k]$.

Figure 7.5 illustrates the dynamics of the system for $K = 0, \cdots , 30$. The six plots indicate two interesting facts, namely the impact of different decision rules in distributed decision-making and the dependence of initial conditions. Note that this dynamic emerges from the internal structure of the CPS Φ.

A detailed study of the different dynamics of CPSs will be presented in the next chapter. Despite the simplicity of the proposed example, it clearly shows the potential of the proposed approach to CPSs as self-developing reflexive–active systems constituted by three layers and cross-layer processes. In other words, the (observable) behavior of this CPS cannot be reduced by the characterization of, for example, one layer or one process alone. The importance of communications between the cyber elements of the system is also clear. Any communication, in its turn, requires some sort of protocol so transmitters and receivers could actually exchange data [13]. A high-level vision of communication protocols and other types of protocols needed to design CPSs will be discussed in the following.

7.6 Layer-Based Protocols and Cyber-Physical Systems Design

The existence of CPSs presupposes a shared symbolic domain where data are structured and meaningful among relevant elements forming then a *common language* with syntax and semantics. Therefrom, messages with semantic value can be mapped into physical communication links through which such elements can exchange data. This, of course, indicates that such a common language exists and is materialized through data processes including transmission and

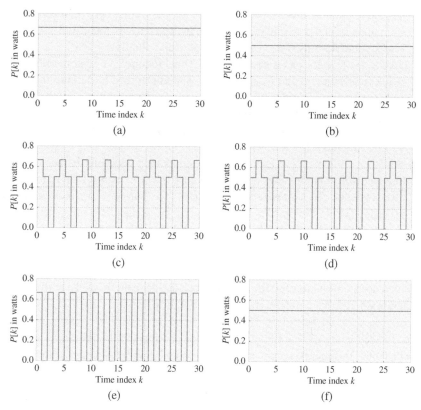

Figure 7.5 Self-developing electric circuit: impact of decision rules and initial conditions. (a) $s_B[k+1] = s_C[k]$ and $s_C[k+1] = s_B[k]$ with $s_B[0] = s_C[0] = 1$. (b) $s_B[k+1] = s_C[k]$ and $s_C[k+1] = s_B[k]$ with $s_B[0] = 0$ and $s_C[0] = 1$. (c) $s_B[k+1] = $ not $s_C[k]$, and $s_C[k+1] = s_B[k]$ with $s_B[0] = s_C[0] = 1$. (d) $s_B[k+1] = $ not $s_C[k]$, and $s_C[k+1] = s_B[k]$ with $s_B[0] = 0$ and $s_C[0] = 1$. (e) $s_B[k+1] = $ not $s_C[k]$, and $s_C[k+1] = $ not $s_B[k]$ with $s_B[0] = s_C[0] = 1$. (f) $s_B[k+1] = $ not $s_C[k]$, and $s_C[k+1] = $ not $s_B[k]$ with $s_B[0] = 0$ and $s_C[0] = 1$.

reception of meaningful physical signals. The study of languages is the focus of specific disciplines, such as linguistics, semiotics, and logic, which have their own (heated) debates and extremely rich findings. Although those domains should be acknowledged, their extensive study of language as a social phenomenon is well beyond our aims; the reader is referred to [14] as a broad overview of the topic.

This book mainly focuses on data generated by machines and communication between them. In a nutshell, the processes at the physical layer need to be converted into (analog or digital) signals, which will be the basis of the data to be further processed, exchanged, and used to (directly or indirectly) intervene in the physical layer. Hence, from the data acquisition phase to its use in decision-making and acting processes, languages specially designed for machines are needed in order to be functional for their specific purposes. Despite the challenges in their design, we factually assume that these languages do exist as part of the enabling ICTs to be presented later in Chapter 9.

This prelude indicates how the physical layer is apprehended and symbolically manipulated through machine-type languages to enable the deployment of CPSs. However, one important missing aspect refers to the way the different elements of the CPS coordinate their actions, and for this, we need *protocols*. This term has been particularly employed in the computer networking literature with the following definition [15]:

> [A] **protocol** defines the format and the order of messages exchanged between two or more communicating entities, as well as the actions taken on the transmission and/or receipt of a message or other event.

We will employ a similar but more general definition as stated next.

Definition 7.2 *Protocols in CPSs* Cyber elements of CPSs are individually controlled by algorithms, which are step-by-step procedures. Protocol is an explicit set of instructional data that are employed to coordinate processes in time and space that involve more than one cyber element of a given CPS. CPS protocols can then be defined within the same layer and across different layers.

Thus, the protocols are designed to guarantee the conditions of reproduction of the CPS in the sense defined in Chapter 2. Each particular CPS requires specific protocols that are codetermined by the intricate intra- and cross-layer interactions. At the same time, the same protocols also codetermine how those interactions are coordinated. A proper conceptualization of any CPS then necessarily requires the determination of its protocols and how they affect, and are affected by, its behavior. Some key aspects of layer-based protocols and how they impact on the design of CPS are summarized below.

- Observation protocol for measuring and sensing cross-layer processes that determine the data acquisition from the physical layer to the data layer. The main factors to be taken into account are: (i) when a new measurement is taken (e.g. periodic sampling, aperiodic event-triggered acquisition, or hybrid between both), and (ii) how the sample is processed to be recorded (e.g. the instantaneous measured value, or the mean value between the two recording times, time-indexed or not, analogue or digital, and how data are coded to be interpretable).

- Protocols for data processes that are employed to produce potentially informative data used by decision-makers. The main factors are: (i) what data are to be used and how to aggregate and fuse them (e.g. based on spatial proximity of sensors, time indices, or statistical relations), (ii) how to solve problems of heterogeneity of sources that may use different codes to record their data (i.e. their interoperability), and (iii) how to preserve secrecy or privacy of data to be shared (e.g. sharing only part of the data, cryptography, or manipulation of metadata).

- Communication protocols that are used to guarantee that the cyber elements can exchange data that will materially enable the structure of awareness of the CPS. This is related to different aspects of communication systems like (i) how to access the physical medium (e.g. wireless or wired), (ii) how to guarantee that the message was successfully received, and (iii) how to send messages over a network. Actually, modern communication networks are also CPSs and can be studied accordingly.

- Decision-making protocols are related to how decisions are made, being them centralized, decentralized, or distributed as discussed in more detail in Chapter 6. The protocol indicates, for instance, when decisions are taken, the order of the decisions in systems with more than one decision-maker, how distributed decision are taken (e.g. majority vote, consensus, logical gates), and the hierarchy of decision-makers (e.g. if there are decision-makers with greater autonomy than others in decentralized systems).

- Action protocols define how and when agents intervene in the CPS. After the information about the decision-making process is available, the protocol defines how to coordinate and organize the agents operating in the CPS to avoid, for example, undesirable collective effects.

More details can be found in the rich literature of different fields such as communication and computing networks, control theory, signal processing, data fusion, and computer sciences [13, 15–17].

Let us return to Example 7.2 in order to illustrate the layer-based protocols and their impact on the CPS behavior.

Example 7.3 *Impact of layer-based protocols in the self-developing electric circuit* Example 7.2 assumed the following protocols.

- Periodic data acquisition with the period Δt of the instantaneous power dissipated by the resistors with synchronous measurements. All measurements are labeled by their time index k and their respective cyber element A, B, and C, being both metadata. The data are recorded as real numbers.

- The protocol for the data process refers to how the element D sums the data acquired by A, B, and C. This operation happens once per acquisition period, and thus, it is also time-indexed by k. Identification of the source node is not needed. The computation time is neglected in comparison with the acquisition period Δt.

- The communication system is ideal so that the transmissions from A, B, and C to D, and between B and C are assumed orthogonal (not interfering with each other); this can be realized by several ways, e.g. by time or frequency division [13]. The transmissions occur once per observation period. We assume that all transmissions are successful and with a negligible latency in comparison with Δt. The metadata in this case is necessarily the identification of the message receiver, and not necessarily the time index k and the sender identification.
- The decision-making is individually carried out by B and C once per observation period and the computation time is negligible compared with Δt. The decision-making process is rule-based and happens after they receive the information of the state of the other element.
- After the decision is taken, B and C (which are both decision-makers and agents) intervene in the physical system by synchronously switching on or off the respective circuit line just before a new measurement is taken. The action time is considered negligible.

It is obvious that this is an ideal CPS defined as a pedagogical example that can provide a trustworthy representation of its observable behavior. If more realistic assumptions are introduced, the protocol should be redesigned. As an example, let us see the impact of a single modification in the action protocol and its impact on the operation of the CPS, as well as on the trustworthiness of the observation.

- **Change:** instead of actions being performed synchronously just before a new measurement is taken, the actions are performed as soon as a decision is taken, and thus, B and C may act asynchronously but still within the Δt range.
- **Impact:** the possibility of asynchronous actions implies that the reported observable behavior $P[k]$ does not reflect changes in the system, including situations when one agent has acted and the other has not, and when both have acted but $P[k]$ has not yet been updated to its new state.
- **Possible solution:** The data acquisition protocol needs to be triggered by events [16] that indicate that a variation in the system state has occurred. The communication and data processing protocols need to be redesigned so that the nonperiodic nature of event-driven acquisition is taken into consideration. The decision-making protocol can be maintained as it is. This system design might result in unstable behavior (not similar to any presented in Figure 7.5) or lead to a stable solution with a fixed value as in Figure 7.5a with B and C on, or B and C off leading to $P[k] = 0$.

Examples 7.2 and 7.3, as well as Section 7.3, indicate that the same CPS may have different behaviors depending not only on the decision rules but also on its initial conditions and protocols. Roughly speaking, one may observe stable, oscillatory, or random dynamics; a theoretically grounded classification will be introduced in the next chapter that focuses on the dynamics of CPSs. It is noteworthy that the proper characterization of the CPS must consider its three layers and cross-layer processes even in the pedagogical examples presented here.

7.7 Summary

This chapter finally introduced the core concept of this book, namely the three layers of CPSs. We indicated how to characterize the physical, data, and decision layers, as well as cross-layer processes. A few idealized examples were provided to illustrate (i) the way the concepts presented in the previous chapters can be employed to characterize CPSs, and (ii) the need for the proposed approach to fully apprehend the system dynamics. In the following chapters, we will build upon this proposed theory to investigate the dynamics of CPSs but still in idealized scenarios (Chapter 8), its main enabling technologies in 2020 (Chapter 9), critical evaluation of existing CPSs (Chapter 10), and considerations beyond technology (Chapter 11). The preliminary studies that resulted in this chapter are reported in the following papers [4–7]. Another approach to CPSs is presented in [16], which is a highly recommended book for advanced readers who are more interested in computer sciences, control theory, and robotics. Although its approach differs from the one taken in this book, the contents surely supplement each other.

Exercises

7.1 Information of events   Consider the numerical example presented in Section 7.3. The task is to characterize the uncertainty of events that generate the outcome observable $P[k]$.
   (a) Define a random variable Y that characterizes the events of interest, i.e. the combination of the switches that are on and off.
   (b) Compute the conditional probabilities $P(Y|X)$ and the conditional entropy $H(Y|X)$.
   (c) Compute the mutual information between X and Y. What does it mean?
   (d) Prove that $H(Y) = H_{\max,Y}$. What is the fundamental assumption that results in this equality?

7.2 Charging electric vehicles   Consider a charging station for electric vehicles (EVs). There are three plugs but only one can be in use while the other two are off because of the capacity limitations of the cables. The task is to design a CPS deployment to coordinate the resource allocation of plugs.

(a) Propose a solution using a centralized decision-making that includes the SAw, SAc, and layer-based protocols.

(b) Propose a solution using a decentralized decision-making that includes the SAw, SAc, and layer-based protocols.

(c) Propose a solution using a distributed decision-making that includes the SAw, SAc, and layer-based protocols.

(d) Compare these three approaches by indicating their benefits and drawbacks.

References

1 Jasperneite J, Sauter T, Wollschlaeger M. Why we need automation models: handling complexity in industry 4.0 and the internet of things. IEEE Industrial Electronics Magazine. 2020;14(1):29–40.

2 Karnouskos S, Leitao P, Ribeiro L, Colombo AW. Industrial agents as a key enabler for realizing industrial cyber-physical systems: multiagent systems entering industry 4.0. IEEE Industrial Electronics Magazine. 2020;14(3):18–32.

3 Uslar M, Rohjans S, Neureiter C, Pröstl Andrén F, Velasquez J, Steinbrink C, et al. Applying the smart grid architecture model for designing and validating system-of-systems in the power and energy domain: a European perspective. Energies. 2019;12(2):258.

4 Kühnlenz F, Nardelli PHJ. Dynamics of complex systems built as coupled physical, communication and decision layers. PLoS One. 2016;11(1):e0145135.

5 Kühnlenz F, Nardelli PHJ, Alves H. Demand control management in microgrids: the impact of different policies and communication network topologies. IEEE Systems Journal. 2018;12(4):3577–3584.

6 Nardelli PHJ, Kühnlenz F. Why smart appliances may result in a stupid grid: examining the layers of the sociotechnical systems. IEEE Systems, Man, and Cybernetics Magazine. 2018;4(4):21–27.

7 Gutierrez-Rojas D, Ullah M, Christou IT, Almeida G, Nardelli PHJ, Carrillo D, et al. Three-layer approach to detect anomalies in industrial environments based on machine learning. In: 2020 IEEE Conference on Industrial Cyber-physical Systems (ICPS). vol. 1. IEEE; 2020. p. 250–256.

8 Cacciapuoti AS, Caleffi M, Tafuri F, Cataliotti FS, Gherardini S, Bianchi G. Quantum internet: networking challenges in distributed quantum computing. IEEE Network. 2019;34(1):137–143.

9 Adonias GL, Yastrebova A, Barros MT, Koucheryavy Y, Cleary F, Balasubramaniam S. Utilizing neurons for digital logic circuits: a molecular communications analysis. IEEE Transactions on Nanobioscience. 2020;19(2):224–236.

10 Lefebvre V. Conflicting Structures. Leaf & Oaks Publishers; 2015.

11 Gerovitch S. From Newspeak to Cyberspeak: A History of Soviet Cybernetics. MIT Press; 2004.

12 Lepskiy V. Evolution of cybernetics: philosophical and methodological analysis. Kybernetes. 2018;47(2):249–261.

13 Popovski P. Wireless Connectivity: An Intuitive and Fundamental Guide. John Wiley & Sons; 2020.

14 Hamawand Z. Modern Schools of Linguistic Thought: A Crash Course. Springer Nature; 2020.

15 Kurose JF, Ross KW. Computer Networking: A Top-Down Approach. Pearson Education, Inc.; 2017.

16 Alur R. Principles of Cyber-Physical Systems. MIT Press; 2015.

17 Lahat D, Adali T, Jutten C. Multimodal data fusion: an overview of methods, challenges, and prospects. Proceedings of the IEEE. 2015;103(9):1449–1477.

8

Dynamics of Cyber-Physical Systems

This chapter covers key aspects of the dynamics of cyber-physical systems (CPSs). We will first indicate different paradigms to characterize CPSs including differential and difference equations, stochastic processes, and an agent-based model. Our aim is to provide a simple overview of different approaches without focusing on the peculiarities of dedicated disciplines like control theory or multiagent systems. As in previous chapters, we rather prefer to study an idealized toy model that is capable of illustrating the main features of the dynamical behavior of CPSs for pedagogical reasons. The exposition will be constructed using a simple elementary cellular automaton (CA) – extensively studied in [1]. By using this idealized model, it will be possible to understand how four classes of behavior can emerge as a result of the internal constitution of the CPS. Possible evaluation metrics and the impact of attacks against the CPS will also be studied following the proposed example.

8.1 Introduction

There are several ways to evaluate the dynamics of different systems, some already presented in previous chapters. The mathematical characterization of dynamical systems using differential equations is probably the best-known approach in engineering because of the physical laws of classical mechanics, thermodynamics, and electromagnetism. Difference equations, which are roughly speaking a discrete version of differential equations, are also frequently employed to study dynamics of systems that are defined by discrete time indices (e.g. population dynamics or traffic models). These methods could be used to study either deterministic or stochastic systems. As indicated in the previous chapter, CPSs assume the existence of decision-makers and agents that usually cannot be explicitly included in simple mathematical equations, and thus, computational approaches might be more suitable. In the following, we will provide a simple overview of the aforementioned methods.

Cyber-physical Systems: Theory, Methodology, and Applications, First Edition. Pedro H. J. Nardelli.
© 2022 The Institute of Electrical and Electronics Engineers, Inc. Published 2022 by John Wiley & Sons, Inc.

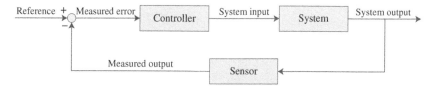

Figure 8.1 Typical block diagram of (negative) feedback control.

The theoretical basis of automatically controlling dynamical systems is the feedback loop [2]. Figure 8.1 presents its typical representation that consists of a reference signal that is compared with the measured output obtained by a sensor, which will lead to a measured error that will be used as the input of a controller that will act to modify the physical system, whose output is measured by the above-mentioned sensor. The usual goal of the control mechanism is to minimize the difference between the reference signal (input) and the measured output. If the measured output is a trustworthy representation of the actual state of the physical system and the measured error tends to zero, then the mechanism designed to control the output of the physical system by a reference signal is successful.

Following the concepts introduced in the previous chapter, Figure 8.1 is a special case of a CPS, which is the scientific object of control theory [3] that has been supporting the technological development of a huge number of devices and tools, such as ovens, airplanes, power grids, and microelectronic circuits. However, our conceptualization is broader, and our study cannot be reduced to control theory despite the recent results in the field of networked control [4].

Before we move ahead, it is important to take a small step back and quickly overview the very basics of a field traditionally called *Signals and Systems* [5]. In its simplest form, continuous-time signals are functions $x : \mathbb{R} \to \mathbb{R}$. For example, $x(t) = 10\sin(2\pi t)\exp(-\pi t/5)u(t)$, where $u(t)$ is the step function defined as $u(t) = 1$ if $t \geq 0$, and $u(t) = 0$ if $t < 0$. In this case, $x(t)$ is a function of the continuous time t. Figure 8.2 illustrates this signal.

This signal can serve as an input of another element, usually denominated in the literature as a *system* (note that his definition of system is different from the one proposed in Chapter 2). Figure 8.3 presents a simple box diagram of the input–output relation caused by a given system. This abstraction represents actual physical relations that can be mathematically represented by operations like derivative, integration, and convolution with other signals. The resistor–capacitor (RC) circuit presented in Figure 8.4 is an example of a system that is mathematically characterized by the following equation

$$x(t) = RC\frac{\mathrm{d}}{\mathrm{d}t}y(t) + y(t),$$

where $x(t) = V_{\mathrm{in}}(t)$ and the output $y(t) = V_{\mathrm{C}}(t)$.

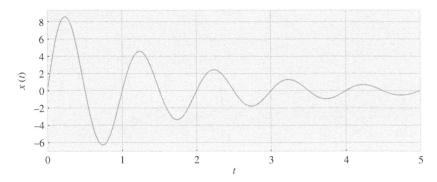

Figure 8.2 Example of a continuous-time signal $x(t)$.

Figure 8.3 Schematic of a system L whose input signal is $x(t)$ and output signal $y(t)$.

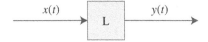

Figure 8.4 RC circuit with an input signal $x(t) = V_{in}(t)$ and an output $y(t) = V_C(t)$; both are measured in volts. The system is defined by a connection between the resistor with the resistance R and the capacitor with the capacitance C.

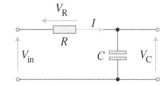

It important to reinforce that this relation is physical, and thus, the goal is to characterize the input–output relation by solving the differential equation. There are different possible ways to do it, but the method using Laplace or Fourier transforms is possibly the most commonly used one. The idea is to map the problem into another domain to then characterize the system by its response to an impulse signal, defining the transfer function of the system. This helps to solve the differential equation of linear and time invariant systems for an arbitrary input signal that has a well-defined Laplace transform. Textbooks like [5] provide all the theoretical background, which is not our focus here, including an extensive analysis of definitions, properties, and classifications used in the field of signals and systems (with a special focus on the linear time-invariant system). Figure 8.5 exemplifies the response $y(t)$ that the RC gives to an input $x(t)$.

A similar conceptualization can be carried out for signals that are discrete in time, i.e. the time is indexed by a variable $k \in \mathbb{Z}$, forming an input sequence (or time series) $x[k]$ and an output sequence $y[k]$. A discrete-time version of the signal presented in Figure 8.2 is shown in Figure 8.6. The relation between input and output is given by difference equations, and the transfer function is obtained from

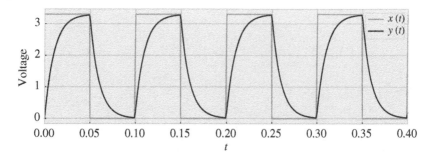

Figure 8.5 Example of the input and output signals in the RC circuit presented in Figure 8.4 with $R = 100\,k\Omega$ and $C = 100\,nF$. The input is a sequence of periodic pulses whose period is 0.1 second.

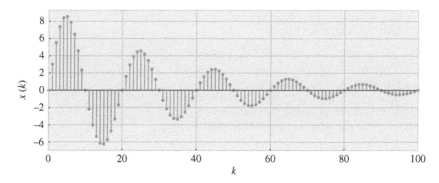

Figure 8.6 Example of a discrete-time signal $x[k]$.

a Z-transform. The above-mentioned book [5] also covers discrete-time signals and systems.

An important distinction worth mentioning is found between periodic and aperiodic discrete-time signals, which is usually associated with event-driven or event-triggered approaches of data acquisition, signal processing, and control [6]. The idea behind this approach is to predefine events that will trigger the acquisition of a new sample or a control action. The events are generally defined through thresholds based on rules such as *acquire a new sample if the measured signal is greater than a given value* and *act in the system if its measured output signal is below a given lower limit*. These rule-based behaviors are usually more complicated to be mathematically characterized (although possible in some cases), and therefore, computational models and heuristics are usually employed.

Another important classification relates to how many elements can take actions within the system boundaries, defining single-agent and multiagent systems. The name multiagent system is quite informative because it is defined as a system composed of two or more elements that can internally take action capable of modifying its dynamics. The effects of each one of the agents and their combined actions in the system depend not only on the physical system itself but also on its structures of awareness and action, as indicated by the examples presented in the previous chapter. In the literature, multiagent systems are usually associated with collaborative control [7] by studying how the measurable attributes of the system are coupled (either via differential equations or computational models). There are a wide range of interesting examples, such as distributed control in power grids [8], swarm robotics [9], and random access in wireless networks [10].

Each case has its own very specific challenges, but all are CPSs following the approach taken in this book. In the next section, we will focus on one abstract example as a pedagogical tool to highlight the most relevant aspects of the dynamics of CPSs that are usually unclear when studying particular cases.

8.2 Dynamics of Cyber-Physical Systems

As extensively discussed in Chapter 7, CPSs can be classified as a subclass of self-developing reflexive–active systems constituted by three layers and cross-layer processes. All CPSs then have the potential to change their internal states and their behavior following their self-development, which may also include internal and external sources of uncertainty, and the relation to the environment as discussed in Chapter 2. A given system is usually classified by the characterization of the dynamics of some of its observable attributes, or metrics derived therefrom.

In this chapter, instead of focusing on any existing system, a simple – but extremely rich in its spatiotemporal dynamics – computational model called an elementary CA (see [1] will be employed as part of the data and decision layers of the CPS to be studied here. A brief description of such a model will be presented next.

8.2.1 Elementary Cellular Automaton

A CA is a symbolic object that is defined as [11]

> (...) discrete, abstract computational systems that have proved useful both as general models of complexity and as more specific representations of non-linear dynamics in a variety of scientific fields. Firstly, CA are (typically) spatially and temporally discrete: they are composed of a finite or denumerable set of homogenous, simple units, the atoms or cells. At each time unit, the cells instantiate one of a finite set of states. They evolve in parallel at discrete time steps, following state update functions or dynamical

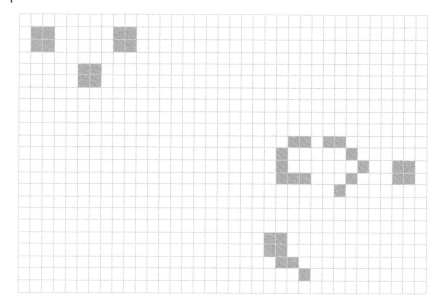

Figure 8.7 Example of a two-dimensional cellular automaton. Black cells are on, while state cells are off. Source: Adapted from https://commons.wikimedia.org/wiki/File:Game_of_life_Simkin_glider_gun.svg.

transition rules: the update of a cell state obtains by taking into account the states of cells in its local neighborhood (there are, therefore, no actions at a distance). Secondly, CA are abstract: they can be specified in purely mathematical terms and physical structures can implement them. Thirdly, CA are computational systems: they can compute functions and solve algorithmic problems.

Figure 8.7 illustrates a snapshot of an example of a two-dimensional CA where the cells can only assume two states, which are either on (black) or off (white).

A very simple class – which is called an elementary CA – is the one-dimensional CA, where cells can be only on or off. The state of each cell depends on a given update rule that depends on the state of the cell itself and the state of its two neighbors, one on the right, the other on the left. Despite its simplicity, several interesting results can be derived from it so much so that Wolfram used it to claim the appearance of a *New Kind of Science* [1]. Without touching his extremely questionable position, Wolfram presents an extensive study of how the spatiotemporal development of the elementary CA may result in different patterns depending on the particular updating rule used by the cells. Before going into these details, we will present the fundamentals of the elementary CA.

Table 8.1 Update rule for the elementary CA.

$s_{i-1}[k]$	$s_i[k]$	$s_{i+1}[k]$	$s_i[k+1]$
1	1	1	c_7
1	1	0	c_6
1	0	1	c_5
1	0	0	c_4
0	1	1	c_3
0	1	0	c_2
0	0	1	c_1
0	0	0	c_0

Definition 8.1 *Elementary CA* The elementary CA is defined as a one-dimensional grid composed of N cells, which can be white representing the state off (or "0") or black representing the state on (or "1"). The temporal development of the cells' states occurs in discrete time. The state of each cell $i \in \{1, \cdots, N\}$ at $k \in \mathbb{N}$ is $s_i[k]$. The update rule of each cell i is a function of the state of its immediate neighbor $i - 1$ and $i + 1$; for the borders $i = 1$ and $i = N$, it is either assumed that the state of the missing neighbor is constant or that such cells are neighbors. Mathematically, we have $s_i[k+1] = f(s_{i-1}[k], s_i[k], s_{i+1}[k])$ where $f : \{0,1\}^3 \to \{0,1\}$, which is represented by Table 8.1.

The variable c_j with $j = 0, \cdots, 7$ is binary, and thus, there are $2^8 = 256$ possible update rules for the elementary CA. Each binary string $c_7 \cdots c_0$ can be seen as a binary number that uniquely identifies the update rule. The binary number $c_7 \cdots c_0$ is presented on the decimal basis, and thus, we have the number of the update rule. The rule number is then computed as $\sum_{j=0}^{7} 2^j c_j$.

Figure 8.8 exemplifies a typical development of an elementary CA, in this case using rule 30 and $N = 31$. Each row of the grid represents the state $s_i[k]$ at a given discrete time k, starting from $k = 0$. Hence, the two-dimensional grid depicts the spatiotemporal development of the elementary CA. For rule 30 with the initial condition $s_i[0] = 0$ for $i \neq 15$ and $s_i[0] = 1$ for $i = 15$, we can see an interesting pattern emerging over time.

Different rules lead to different patterns, as indicated by Figure 8.9. Wolfram identified four different classes of patterns [pp. 231–235][1]:

In class 1, the behavior is very simple, and almost all initial conditions lead to exactly the same uniform final state.

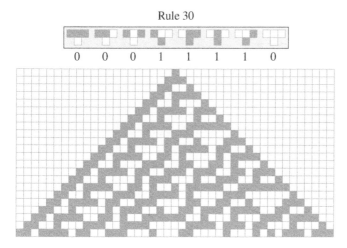

Figure 8.8 Temporal development of an elementary CA with rule 30 for $N = 31$ and initial states $s_i[0] = 0$ for $i \neq 15$ and $s_i[0] = 1$ for $i = 15$, considering the border nodes $i = 1$ and $i = 31$ assuming that their missing neighbors are in state 0 (white). Source: Adapted from https://mathworld.wolfram.com/Rule30.html.

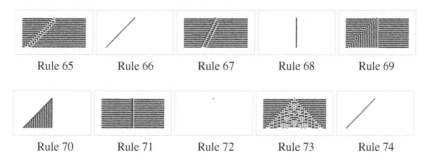

Figure 8.9 Development of an elementary CA with different rules (from 65 to 74); different spatiotemporal patterns are visually noticeable. Source: Adapted from https://en .wikipedia.org/wiki/Elementary_cellular_automaton.

In class 2, there are many different possible final states, but all of them consist just of a certain set of simple structures that either remain the same forever or repeat every few steps.

In class 3, the behavior is more complicated, and seems in many respects random, although triangles and other small-scale structures are essentially always at some level seen.

(…), class 4 involves a mixture of order and randomness; localized structures are produced which on their own are fairly simple, but these structures move around and interact with each other in very complicated ways.

Of course, the development of the CA is deterministic given the set of initial conditions $s_i[0]$. The statistical analysis used by Wolfram to classify the spatiotemporal pattern generated by each rule, as well as other forms of classification, are beyond our scope here. The visual appeal is possibly the key here.

What is important for us is to know that the CA self-development is associated with the updating rules and the initial conditions, whose spatiotemporal dynamics in the long run consists of:

- a uniform pattern (class 1);
- a periodic pattern (class 2);
- chaotic (aperiodic) patterns (class 3);
- complex patterns with localized structures (class 4).

In the next section, we will study how the elementary CA could serve to represent the data and decision layers of a CPS.

8.2.2 Example of a Cyber-Physical System

Consider a CPS in which the physical layer is the electric circuit presented in Figure 8.10. There is a constant voltage source V associated with a resistor R_V that supplies electric power to N pairs of resistors in parallel, which may represent a toy model of microgrids [12]. Each pair has one resistor R_{bi} that is always active (i.e. a base load) and another R_{fi} that may be connected or not depending on the state on ($s_i = 1$) or off ($s_i = 0$) of its associated switch (i.e. flexible load).

If we assume that $R_{bi} = R_{fi} = R, \forall i \in \{1, \cdots, N\}$ and that the number of resistors in the on state is $h = \sum_{i=1}^{N} s_i$, then we can compute the equivalent resistor of the circuit R_{eq} as

$$R_{eq} = R_V + \frac{R}{N+h}. \tag{8.1}$$

The extreme cases are then when all the flexible loads are connected, and we have thus $R_{eq}^{all} = R_V + R/N$ and none $R_{eq}^{none} = R_V + R/2N$. If different discrete observation times $k \in \mathbb{N}$ are considered, then we have $s_i[k]$, $h[k]$ and $R_{eq}[k]$.

Now, this circuit is the physical layer of CPS Φ that is a discrete-time self-developing system where its data and decision layers are constructed upon the elementary CA with N elements. Each element $A_i \in \{1, \cdots, N\}$ is both a decision-maker that follows the updating rule defined by the CA and an agent to connect or disconnect the flexible load accordingly. We can then define Φ as follows.

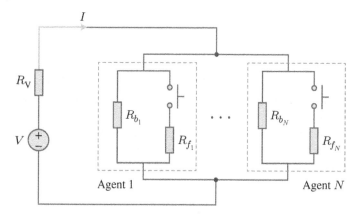

Figure 8.10 Example of an electric circuit with *N* agents. Source: Adapted from [12, 13].

(1) **Input:** V in volts, and **output:** $P[k] = V^2/R_{eq}[k]$ in watts where the equivalent resistor $R_{eq}[k]$ in ohms depends on the number $h[k]$ of how many elements are states (on or off) at discrete-time k.

(2) **SAw:** $1 + \sum_{i=1}^{N} a_i + a_1(a_1 + a_2) + \sum_{i=2}^{N-1} a_i(a_{i-1} + a_i + a_{i+1}) + a_N(a_{N-1} + a_N)$. This indicates that all elements receive data directly from the physical layer. Besides, the level 2 processes indicate that to make a decision, the element A_i has an image of its own state as well as of its direct neighbors A_{i-1} and A_{i+1}, also considering that the border elements have only one neighbor.

(3) **SAc:** $\sum_{i=1}^{N} a_i$. This indicates that elements A_i are agents that directly act on their own switches.

Remember that the SAw tells nothing about the trustworthiness of the data. Additionally, in this specific case, the SAw and SAc are fixed over time.

Figure 8.11 presents the outcome of the CPS whose input is V and the output is $P[k = \tau]$ for a given $h[k = \tau]$ where $\tau \in \mathbb{N}$ represents a specific discrete time index; the resistors are arbitrarily chosen as $R_V = 0.1$ mΩ and $R = 1\ \Omega$. This figure presents the dissipated power considering different numbers of active flexible loads, which are directly obtained from the state of the CA in time $k = \tau$. It is important to note that we are analyzing this CPS as a "black box" where, for a given input V, we can only observe the total dissipated power $P[k]$ as the output of the discrete time k. The relation between the CPS internal dynamics and the observable outcomes will be presented next.

8.2.3 Observable Attributes and Performance Metrics

As indicated in the previous section, the only attribute of the system that can be observed by an external element is the dissipated power $P[k = \tau]$ at time $k = \tau$ by the CPS Φ for a given input voltage V. If the observable variable $P[k = \tau]$ is

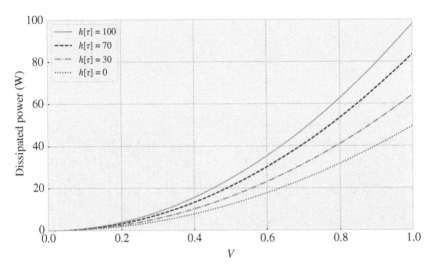

Figure 8.11 Numerical example of the CPS Φ for $R_V = 0.1$ mΩ and $R = 1\$ \Omega$. The input is V in volts and the output is the total power $P[k = \tau]$ in watts dissipated in time $k = \tau$ considering different numbers $h[k = \tau]$ of active flexible loads.

trustworthy, then it is possible to unambiguously determine the number of flexible loads $h[k = \tau]$ that are active in time $k = \tau$. However, the sequences $P[k]$ and $h[k]$ could potentially be the outcome of different rules of the elementary CA.

Depending on the CPS requirements, even though a given outcome $P[k]$ might be acceptable, it may also be produced by an internally undesirable dynamics. For example, a similar $P[k]$ behavior might be produced by a fair activity alloca-tion. This can be measured by a simple performance metric that computes the ratio between how many times a given agent i was in an active state, i.e. $s_i[k = \tau] = 1$, and the time window under consideration. Mathematically, we have $r_i = \sum_{\tau=0}^{\tau_{max}} s_i[k = \tau]/(\tau_{max} + 1)$, considering an arbitrary time window of $\tau_{max} + 1$ start-ing at $k = 0$ and ending at $k = \tau_{max}$. At the system level, another performance metric could be the ratio between the number of agents in the active state $h[k = \tau]$ divided by the number of agents N.

Figure 8.12 illustrates the dynamics of CPS Φ for two different update rules, namely 54 and 73, with a total of $N = 201$ flexible loads controlled by their respective agents. The physical layer setting is $R_V = 0.1$ mΩ and $R = 1$. The input is $V = 1$ V (fixed) and the output is the observable sequence $P[k]$, whose values depend on the initial states and the aforementioned update rules. For this experiment, a random initial state is considered where each state $s_i[0]$ with $i \in \{0, \cdots, 200\}$ is randomly chosen following an independent and identically distributed random variable where $P(s_i[0] = 0) = 0.7$ and $P(s_i[0] = 1) = 0.3$. We set the same initial states for both cases, and the difference in their dynamics is

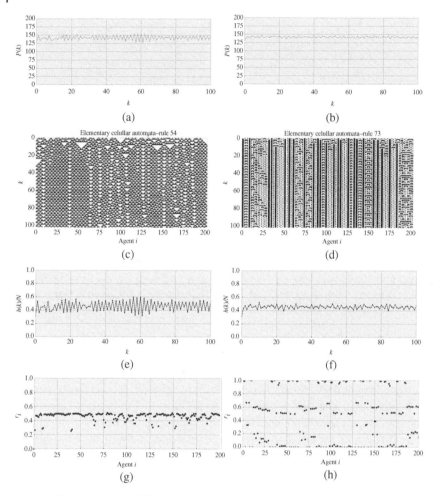

Figure 8.12 Dynamics of CPS Φ for rules 54 and 73 considering $V = 1\,\text{V}$, $R_V = 0.1\,\text{m}\Omega$, and $R = 1\,\Omega$. (a) Sequence $P[k]$ (observable variable) for rule 54. (b) Sequence $P[k]$ (observable variable) for rule 73. (c) CA development for rule 54. (d) CA development for rule 73. (e) Ratio $h[k]/N$ for rule 54. (f) Ratio $h[k]/N$ for rule 73. (g) Ratio r_i considering agent $i \in \{0, \cdots, 200\}$ for rule 54. (h) Ratio r_i considering agent $i \in \{0, \cdots, 200\}$ for rule 73.

only due to the decision rules based on the CA. We study a time window starting from $k = 0$ and ending in $k = 100$.

There are interesting things to note in this illustrative scenario, which we will list below.

- Figure 8.12a, b indicate that the two CPSs lead to sequences $P[k]$ with a similar mean value of dissipated power (approximately 150 W) and a similar behavior, although the first case presents more oscillations.
- Rules 54 and 73 have different classes as presented in Figure 8.12c and d, respectively. Rule 54 is class 4 (complex patterns, visually identified by the triangles of different sizes), while rule 73 is class 2 (periodic behavior).
- The ratio of active flexible loads in the physical layer is also similar, around 50%, as shown in Figure 8.12e and f (although the first varies more). This is true for both rules regardless of the initial condition being $h[0]/N = 0.3$.
- Figure 8.12g, h show a remarkable difference of fairness related to the different agents (and their flexible loads). The first case (rule 54) has the largest majority of its flexible loads active with a similar ratio $r_i \approx 0.5$ (i.e. activity frequency is around 1-out-of-2). The second case (rule 73) has a very large variation: several flexible loads almost never active $r_i \approx 0$, others almost always active $r_i \approx 1$, others with $r_i \approx 0.5$, and still a few others with different ratios.

What is remarkable is that both CPSs have a similar observable outcome $P[k]$ that is the result of very different internal dynamics. This fact is not always true because different rules may lead to different observable outcomes, as it may be inferred from Figure 8.9. Besides, some rules may be more sensitive to different initial conditions. These aspects will be presented in the next section when we aim to optimize the CPS dynamical behavior.

8.2.4 Optimization

Optimization as described in Chapter 6 is associated with the determination of operational points or system parameters that maximize or minimize a given performance metric subject to a set of constraints. The proposed CPS Φ is an example of a self-developing system such that an optimization would in principle be unfeasible. To formulate a proper optimization problem, we first need to specify its desirable operational outcomes, internal constraints, and design parameters. These specifications are given below.

- The operational objective is to guarantee that $P[k]$ is within the range determined by the upper and lower limits of dissipated power.
- All flexible loads should be active with a similar frequency.
- The parameters of the physical layer are given and fixed, as well as the SAw and the SAc.
- The only design parameter is the update rule of the elementary CA.
- The initial conditions are unknown.

In this case, the optimization problem could be formulated in two different ways as follows:

(1) *select the rule that minimizes the frequency that P[k] is out of its operational range subject to a fair activity frequency among the flexible loads*, or

(2) *select the rule that maximizes the fairness of activity frequency among the flexible loads subject to P[k] within its operational range.*

Although it would be possible to write it both in mathematical terms and possibly solve it at least for special cases, we rather prefer typical outcomes considering different update rules and random initial conditions. Such a numerical analysis is presented in Figure 8.13, where the dynamics of the dissipated power $P[k]$ is depicted considering rules 81, 82, 84, 110, 240, and 250 for three different initial conditions, and $V = 1$ V, $R_V = 0.1$ mΩ, and $R = 1$ Ω. By inspection, rule 110 seems a suitable rule because it is the only one that is consistently within the operational range. It is interesting to note that rule 110 tends to move quickly to the desired range regardless of the initial condition, while the other rules seem to have either different "attractors" or an oscillatory behavior heavily dependent on the initial conditions.

However, this figure does not indicate the fairness in the flexible load activity in the system. Figure 8.14 presents the performance of rule 110 by showing the $P[k]$ dynamics over time for three different initial conditions together with the allocation fairness evaluated by the ratio r_i. Besides, Figure 8.14 also shows the same plots for rule 54, which presented a reasonable performance as indicated by Figure 8.12. Figure 14a confirms that the CPS Φ mostly work within its operational limits. However, Figure 8.14b shows that the fairness of the system it not so high, regardless of the initial condition, because the ratios r_i are quite dispersed (but not as much as in the situation presented in Figure 8.12h). On the other hand, Figures 8.14c and d present a different behavior: a stronger dependence on the initial conditions and a fair activity frequency of flexible loads. It is also interesting to see that, although being always around the operational limits, the CPS Φ rarely operates with the desired range. An interesting note is that both rules 110 and 54 are class 4.

Considering all the rules studied in this section, we infer that rule 110 would be the optimal solution among the options presented here. It seems to provide a system whose internal dynamics will tend to a sequence that operates almost always within the required range regardless of the initial conditions, while the different flexible loads have a reasonably fair distribution. However, this is just an indication, and there might be another rule that provides better outcomes. It would also be interesting to prove mathematically the insights provided by these numerical examples. One remarkable thing is that different classes of rule may have the same observable outcomes but produced by a quite different internal dynamic. These results, however, considered an ideal scenario without any failure or intentional attacks against the CPS Φ, which is the topic of the following section.

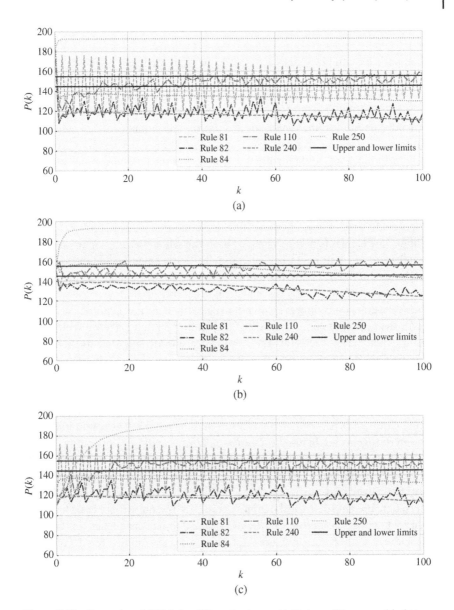

Figure 8.13 Dynamics of CPS Φ for different rules and initial conditions considering $V = 1$ V, $R_V = 0.1$ mΩ, and $R = 1$ Ω. (a) Sequence $P[k]$ (observable variable) for the initial condition uniformly distributed so that $P(s_i[0] = 0) = 0.9$ and $P(s_i[0] = 1) = 0.1$. (b) Sequence $P[k]$ (observable variable) for the initial condition uniformly distributed so that $P(s_i[0] = 0) = 0.5$ and $P(s_i[0] = 1) = 0.5$. (c) CA development for initial condition uniformly distributed so that $P(s_i[0] = 0) = 0.1$ and $P(s_i[0] = 1) = 0.9$.

8.3 Failures and Layer-Based Attacks

In comparison with physical systems, CPSs have an increased vulnerability. The constitution of CPSs in three layers opens new possibilities of failures related to the cyber domain. For example, the data acquired by sensors might be noisy, or communication links might be subject to errors. These types of issues may lead to misinformation (unintentional) or disinformation (intentional), as defined in Chapter 4. Regardless of their nature, untrustworthy data may result in decisions and then actions that would modify the dynamics of the CPS potentially affecting all three layers. Throughout this section, we will analyze the impact of failures and attacks based on the already discussed CPS Φ.

At the physical layer, failures or attacks are related to the electric circuit depicted in Figure 8.10 itself. A wire or cable could be (intentionally or not) broken, disconnecting some elements of the system or even the whole system. There are other possibilities: the input could be (intentionally or not) modified or the switches could be (intentionally or not) broken. In all those cases, the changes in the system dynamics are captured by the mathematical equations.

These modifications, however, do not necessarily alter the acquired data to be used in the cyber domain. For instance, depending on how data are acquired, a broken switch may not act as expected by the decision-making element, and thus, the state of the agent might be different from the actual physical situation. If this is the case, the data will be unrelated to the actual physical state, and thus, the outcome of the CPS will be affected. Besides, this effect also propagates because of the SAw that requires communication between the agents, which also use the problematic data as part of their on decision-making process. The communication between the agents might also be a source of failures in the CPS operation.

Figure 8.15 exemplifies a misinformation cyberattack where the communication link from Agent 99 and Agent 100 is actively attacked by injecting a misinformation of the state of the former, namely $s_{99 \to 100}[k]$, $\forall\, k \geq 20$, to modify the decisions of the latter, affecting then the dynamics of the CPS. By comparing Figure 8.15a–c, it is easy to verify the impact of the cyberattack and its dependence on the fake state, either $s_{99 \to 100}[k] = 0$ in Figure 8.15b or $s_{99 \to 100}[k] = 1$ in Figure 8.15c. Figure 8.15d demonstrates the physical effect of the injection of fake data, modifying the temporal development of the CPS, which is reflected by the change of the observable sequence $P[k]$ for $k \geq 20$. The activity ratio r_i is also affected by the misinformation attack. In this specific case governed by rule 245, the changes only affect the elements on the right side of the attacked one, reaching one more element at each discrete time.

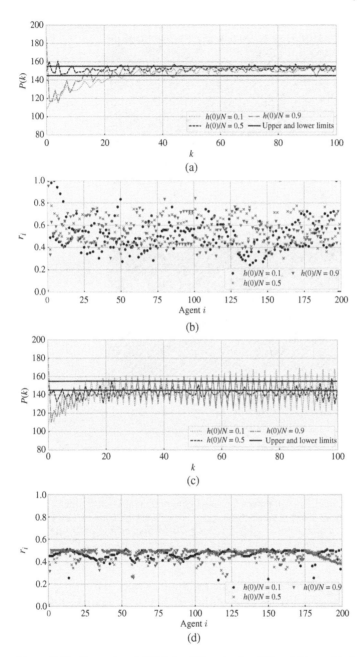

Figure 8.14 Dynamics of CPS Φ for different rules 110 and 54 considering $V = 1$ V, $R_V = 0.1$ mΩ, and $R = 1$ Ω, and different initial conditions. (a) Sequence $P[k]$ for rule 110. (b) Ratio r_i considering agent $i \in \{0, \cdots, 200\}$ for rule 110. (c) Sequence $P[k]$ for rule 54. (d) Ratio r_i considering agent $i \in \{0, \cdots, 200\}$ for rule 54.

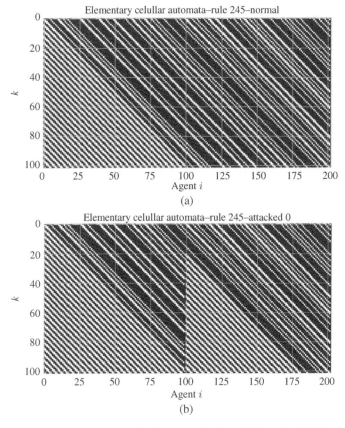

Figure 8.15 Dynamics of CPS Φ for rule 254 considering $V = 1$ V, $R_y = 0.1$ mΩ, and $R = 1$ Ω for the initial condition uniformly distributed so that $P(s_i[0] = 0) = 0.5$ and $P(s_i[0] = 1) = 0.5$. A cyberattack is injected at the communication link from agent 99 to 100 staring at time $k = 20$ so that the latter will always receive a fake state $s_{99 \to 100}[k]$, $\forall k \geq 20$. (a) CA development for rule 245 in normal operation. (b) CA development for rule 245 with a cyberattack $s_{99 \to 100}[k] = 0$, $\forall k \geq 20$. (c) CA development for rule 245 with a cyberattack $s_{99 \to 100}[k] = 1$, $\forall k \geq 20$. (d) Sequence $P[k]$ for rule 245 with and without a cyberattack. (e) Ratio r_i considering agent $i \in \{0, \cdots, 200\}$ for rule 245.

The same procedure could be performed with other rules and other physical layers, and thus, other particular results will be found. What is important to keep in mind is that the three layers are constitutive of CPSs, and thus, all the three layers are vulnerable to attacks. A simple cyberattack may indirectly affect the dynamics of observable physical variables without any foreseeable justification.

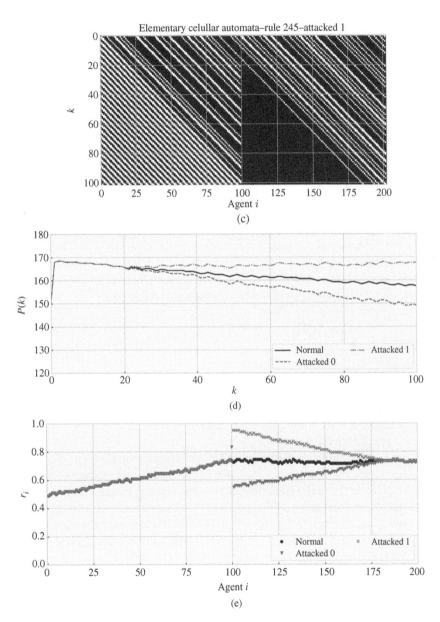

Figure 8.15 (*Continued*)

8.4 Summary

This chapter introduced important ideas related to the dynamics of CPSs following the three-layer approach. Our aim here was to provide an intuitive while theoretically sound example of a CPS whose data and decision layers are defined by the elementary CA, well known for its four classes of spatiotemporal classes of behavior [1]. We have also shown how the system dynamics based on observable variables might hide an intricate internal dynamic. In addition, a new cyber domain enables different sources of failure points and vulnerabilities compared with purely physical systems. Although linked to the theme of this chapter, topics related to established disciplines like systems and signals [5] and control theory [3, 4] were not covered, and the readers are suggested to refer to them by incorporating the theory of CPS presented here. The book [14] focuses on different aspects of the combination of control theory and computer sciences to design CPSs, which can be seen in many ways complementary to the approach focusing on a representative but purified model taken by this chapter. In the next part of this book, we will move from the abstract theory proposed in parts 1 and 2 to then return to aspects concerning the material world, including examples of CPSs, the enabling information and communication technologies, and social impact.

Exercises

8.1 **CPS dynamics**   Consider the CPS analyzed in Sections 8.2 and 8.3. Study four different rules following the examples provided in that section, repeating the results from Figures 8.12 to 8.15. Analyze the results. The code is available at https://github.com/pedrohjn.

8.2 **Impact of defining initial conditions**   Consider the same CPS as the one studied in Exercise 8.1. The aim of this task is to analyze the impact of defining the initial conditions of the elements. Find a combination of an initial condition and rule that provides the desirable outcome based on the specification given in Figure 8.14. *Hint:* Find the desired dynamics of the sequence $P[k]$, and then, write the truth table as presented in Table 8.1 to find the desired rule.

References

1 Wolfram S. A New Kind of Science. vol. 5. Wolfram Media, Champaign, IL; 2002.

2 Wiener N. Cybernetics or Control and Communication in the Animal and the Machine. MIT Press; 2019.

3 Lewis FL. Applied Optimal Control and Estimation. Prentice Hall PTR; 1992.

4 Zhang XM, Han QL, Ge X, Ding D, Ding L, Yue D, et al. Networked control systems: a survey of trends and techniques. IEEE/CAA Journal of Automatica Sinica. 2019;7(1):1–17.

5 Oppenheim AV, Willsky AS, Nawab SH. Signals & Systems (2nd Ed.). USA: Prentice-Hall, Inc.; 1996.

6 Miskowicz M. Event-Based Control and Signal Processing. CRC Press; 2018.

7 Lewis FL, Zhang H, Hengster-Movric K, Das A. Cooperative Control of Multi-Agent Systems: Optimal and Adaptive Design Approaches. Springer Science & Business Media; 2013.

8 Sahoo S, Mishra S, Jha S, Singh B. A cooperative adaptive droop based energy management and optimal voltage regulation scheme for DC microgrids. IEEE Transactions on Industrial Electronics. 2019;67(4):2894–2904.

9 Rinner B, Bettstetter C, Hellwagner H, Weiss S. Multidrone systems: more than the sum of the parts. Computer. 2021;54(5):34–43.

10 Popovski P. Wireless Connectivity: An Intuitive and Fundamental Guide. John Wiley & Sons; 2020.

11 Berto F, Tagliabue J Zalta EN, editor. Cellular Automata. Metaphysics Research Lab, Stanford University; 2021. https://plato.stanford.edu/archives/spr2021/entries/cellular-automata/.

12 Kühnlenz F, Nardelli PHJ, Alves H. Demand control management in micro-grids: the impact of different policies and communication network topologies. IEEE Systems Journal. 2018;12(4):3577–3584.

13 Kühnlenz F, Nardelli PHJ. Dynamics of complex systems built as coupled physical, communication and decision layers. PLoS One. 2016;11(1):e0145135.

14 Alur R. Principles of Cyber-Physical Systems. MIT Press; 2015.

Part III

9

Enabling Information and Communication Technologies

We are now at the beginning of the third part of this book, which will focus on more concrete aspects of cyber-physical systems (CPSs) guided by the theoretical concepts presented so far; this will include data transmission and processing technologies, CPS applications, and their potential social impacts. The focus of this chapter is on the key information and communication technologies (ICTs) that afford the existence of CPSs. Instead of describing in detail the devices, protocols, algorithms, and standards, which would be hardly possible, our choice is to critically review three prominent ICT domains, namely data networks (and the Internet), advanced statistical methods for data processing, and new data storage paradigms. Specifically, we will provide an overview of the key features of the fifth generation (5G) of mobile networks and machine-type wireless communications, the strengths and limitations of machine learning and artificial intelligence (AI), and the advantages and drawbacks of distributed ledgers like blockchains and distributed computing like federated learning. Besides those trends, a speculation of the future of ICT will be presented considering the implications of quantum computing and the Internet of Bio-Nano Things (IoBNT).

9.1 Introduction

The technological development related to ICTs can be dated back to the years after World War II, in a specific conjuncture of the Cold War and Fordism [1]. In the 1960s, there was a great development of the cybernetics movement, firstly in the USA but also in the Soviet Union [2]. One remarkable milestone was the development of the packet switching technology, which allowed multiplexing of digital data messages and was essential to the deployment and scaling up of data networks [3]. This was the beginning of the Internet. Since the nineties, there has been a remarkable increase in deployments of wireless networks with different standards, some dedicated to voice as in the first two generations of cellular

Cyber-physical Systems: Theory, Methodology, and Applications, First Edition. Pedro H. J. Nardelli.
© 2022 The Institute of Electrical and Electronics Engineers, Inc. Published 2022 by John Wiley & Sons, Inc.

networks and others to digital data communication like the ALOHANet and then WiFi. These technologies were only possible because of the huge efforts in the research and development of digital wireless communication technologies [4, 5]. A schematic presentation of the history of telecommunication until the early two thousands can be found in [6]; more recent developments of cellular networks are presented in [7].

Currently, in the year 2021, with the already worldwide established fourth generation (4G) of mobile networks and the first deployments of 5G as well as other types of wireless networks, we are experiencing an undeniable convergence of the Internet and wireless communication including more and more applications. This is evinced, for example, by the steady growth of the Internet of Things (IoT) with protocols dedicated to machine-type communication (MTC), virtual and augmented reality applications, and human-type communications being mostly carried over the Internet and including not only voice but also video (including video conferences). The current discussions about what will be the sixth generation (6G) of cellular networks indicate those and other trends [8].

This is the context in which this book is written. The societal dependence of ICTs is growing at a fast pace and everywhere, even more remarkably after the mobility restrictions imposed during the COVID-19 pandemic. They provide the infrastructure for the data and decision-making layers that are necessary to build CPSs, as conceptualized in the previous two chapters. Note that it is possible and desirable to analyze ICT infrastructures as CPSs, and thus, our approach will consider both physical and logical relations, as well as decision-making processes. In what follows, we will present a brief introduction of data networks with particular attention to wireless systems, mostly focusing on applications related to the IoT and the associated MTCs.

9.2 Data Networks and Wireless Communications

This section provides a brief introduction to data networks highlighting the function that wireless technologies have in their operation. The first step is to follow the framework introduced in Chapter 2, and thus, analyze a data network as a particular system (PS) whose boundaries are defined by its peculiar function (PF), also determining its conditions of existence and main components. This is presented next following the concise introduction from [Ch. 1][3].

Definition 9.1 *Data network as a system.* Data networks refer to an infrastructure that connects physically and logically different devices and components in order to support the operation of different applications. The main components of this PS followed by its PF and conditions of existence (C1–C3) are:

PS (a) Structural components: cables, antennas, connectors, microelectronic devices, microcontrollers, base stations; (b) operating components:

computers, mobile devices, routers, packet switches, servers, signal modulators, communication and data processing protocols; (c) flow components: digital data packets.

PF Interchange of digital data between devices physically connected as a network to serve different end use applications.

C1 It is physically possible to send data packets from an arbitrary point of the network to another coded as digital signals through transmission over potentially different physical media.

C2 Effective protocols to code and encapsulate the data to provide a reliable and efficient communication; existence of data sources and sinks, and thus, communication between points is needed; maintenance of cables, servers, and routers; standards to regulate the use of a shared radio spectrum and cable infrastructure to avoid interference.

C3 Technical training for persons working in the research, development, and maintenance; investments in components needed; agreements in terms of the standardization process; geopolitical relations between countries; protection of the physical infrastructure against malicious attacks and other hazards; availability of electricity (reliable connection to the electricity grid).

Figure 9.1 depicts an example of a data network composed of different devices that run applications sending, requesting, and receiving data (e.g. computers, tablets, mobile phones, electricity metering), data servers (e.g. data centers and super computers, which are usually called *cloud*), and operational components that process data so that they can be transmitted reliably to their final destination (e.g. routers, link layer switches, modems, base stations). These elements are connected forming a decentralized network with some hubs via different physical media, namely cable (fiber-optic and coaxial) and radio (wireless links). The proposed illustration is a simple, but yet representative, subset of the whole Internet.

In the following subsections, a schematic presentation of the Internet will be provided, briefly explaining its layered design (note that this is different from our three layers of CPSs and other layered models) and differentiating the edge and core of the network. Broadly speaking, the objective is to show how logical links between different end devices (which can be located very far away from each other) that exchange data can be physically performed with an acceptable quality of service.

9.2.1 Network Layers and Their Protocols

Data networks and the Internet are designed considering a layered architecture with respect to the services that each layer provides to the system. The advantages of this approach are fairly well described in textbooks like [3, 4]. What interests us here is that the layered approach is an effective way to organize the operation of a potentially large-scale decentralized system, allowing elements and techniques to

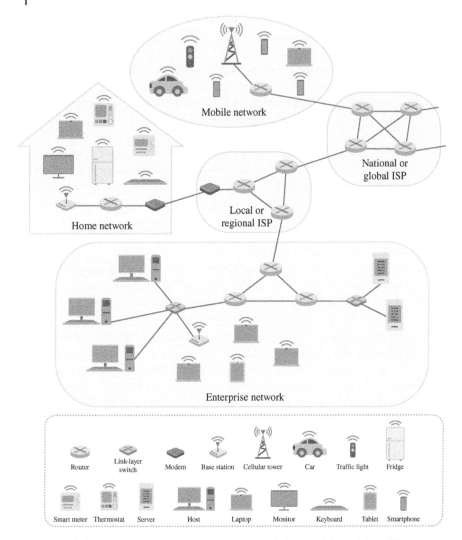

Figure 9.1 Illustration of a small-scale data network. Source: Adapted from [3].

be included and excluded. In the following, we will provide an illustration of how the layered model works based on an example before going into the particularities of data network layers and protocols.

Example 9.1 *Travel from Helsinki to São Paulo.* The distance between Helsinki (Finland) and São Paulo (Brazil) is around 11 000 km, and the most feasible option is to travel by airplane. The national and international air traffic

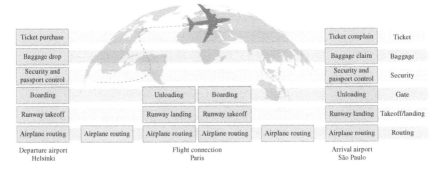

Figure 9.2 Travel from Helsinki to São Paulo via Paris following a layered approach. Source: Adapted from [3].

system is a large-scale decentralized system, which operates in layers. One layer is the ticket layer related to purchase services. Other layer is the airport layer related to departure and arrival services. One final layer is the air traffic itself, where the airplanes fly from one given destination to another. These layers are relatively independent, and thus, the way the ticket is acquired (e.g. by on-line shopping, by phone, or by using a travel agency) makes little difference in the way the departure is organized, or in the actual route selected from one airport to another.

To travel from Helsinki to São Paulo, these general layers need to be followed as exemplified in Figure 9.2. The first step is to acquire the ticket that will give the information related to travel, including the details of the airport layer, including a connection in Paris. By following the instructions contained in the ticket, the person can arrive at the airport and follow the boarding procedures (check in, baggage drop, security check, passport control, and boarding). After the boarding, the pilot flies from Helsinki to Paris following the route proposed by the flight controllers, which is related to the air traffic layer. In Paris, the person needs to follow the new procedures of the airport layer until the boarding to the new flight to São Paulo. After the arrival to the final destination, new procedures of the airport layer are followed (unloading the plane, passport control, and baggage claim). The final step is again in the ticket layer to check if all the information is correct, including the time and place of arrival. If any deviation has happened, the traveler is back to the ticket layer to fill in a complain.

Similar to this example, the operation of data networks and the Internet are layered based on services and functionalities that each layer provides. Particularly, the *Open System Interconnection* (OSI) model was proposed based on seven layers that allow decentralized data transfer, logically connecting end applications through physical (wired and/or wireless) connections. A systematic presentation of the OSI model is given in Figure 9.3.

Layer		Protocol data unit (PDU)	Function
Host layers	7 Application	Data	High-level APIs, including resource sharing, remote file access
	6 Presentation		Translation of data between a networking service and an application; including character encoding, data compression and encryption/decryption
	5 Session		Managing communication sessions, i.e., continuous exchange of information in the form of multiple back-and-forth transmissions between two nodes
	4 Transport	Segment, Datagram	Reliable transmission of data segments between points on a network, including segmentation, acknowledgement and multiplexing
Media layers	3 Network	Packet	Structuring and managing a multi-node network, including addressing, routing and traffic control
	2 Data link	Frame	Reliable transmission of data frames between two nodes connected by a physical layer
	1 Physical	Bit, Symbol	Transmission and reception of raw bit streams over a physical medium

Figure 9.3 Seven layers of the OSI model. Source: Adapted from https://en.wikipedia.org/wiki/OSI_model.

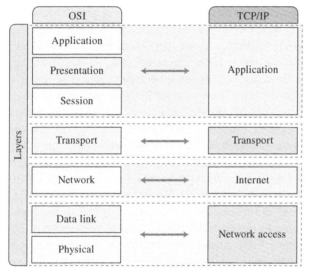

Figure 9.4 OSI reference model and the Internet protocol stack. Source: Adapted from https://upload.wikimedia.org/wikipedia/commons/d/d7/Application_Layer.png.

These seven layers constitute a reference model that guides the actual data network design and deployment, which nevertheless can be (and usually are) modified. For example, the Internet has a different layer structure than the OSI model, although a map between them is possible as illustrated in Figure 9.4. The layered model defines the *protocol stack*, determining the data structure and the interactions allowed at each layer. The Internet is generally constituted by the Internet Protocol (IP) related to the OSI network layer and the Transmission Control Protocol (TCP) related to the OSI transport layer.

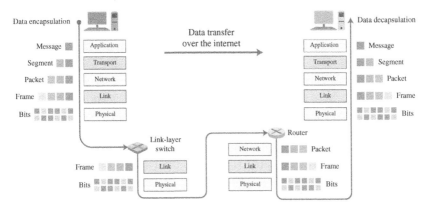

Figure 9.5 Encapsulation of a generic data transfer over the Internet. Source: Adapted from [3].

To accomplish its peculiar function, the data network needs to logically connect two elements, namely a source of data associated with a given application and its respective destination. Examples of applications are email, instant messages, and remote monitoring of sensors. The creation of logical links through physical media requires a process of *encapsulation*, where at each layer a new piece of data is added to accomplish its functionality. Figure 9.5 illustrates how data are transferred over the Internet, including (i) message encapsulation at the source, (ii) physical transmissions, (iii) routing, and (iv) message decapsulation at the destination.

The operation of the data network can be analyzed from two perspectives, one physically closer to the end users (i.e. the source and destination) and the other inside the network where data are routed. The first case refers to the *network edge* and the second to the *network core.* We will discuss them in the following subsections.

9.2.2 Network: Edge and Core

In our daily lives, we constantly interact with and through data networks via devices like personal computers, smart phones, and tablets. These are denominated as *end systems*, which are also called *hosts* in the literature because of their function of hosting and running application programs. There are also other end systems such as all sorts of sensors, smart appliances, and advanced electricity meters. They constitute what we call today the IoT. All those end systems are usually called *clients*.

In addition, there is another class of end systems called *servers* that are associated with large data centers and supercomputers, whose function is to store large amounts of data and/or to perform demanding calculations. These are usually invisible pieces of hardware for the human user that enabled the *cloud computing* paradigm. They are generally located far away from the clients, potentially leading to a poor quality of service in terms of delay and reliability. Nowadays, in 2021, new paradigms are being developed to locate some smaller servers closer to the clients defining the *edge computing* paradigm, or even considering the clients as a potential source of computing power defining the *fog computing* paradigm. These are recent terms and their meaning might vary, although the tendency of decentralization of computing power and distributed computation is captured [9].

The network edge is composed of those end systems and applications that are physically connected to (and through) the core of the data network. Such connections happen through the *access network*, which is the network that exists before the first router (called *edge router*). There are several types of access network, including mobile networks in outdoor environments (e.g. 4G cellular systems), home networks (e.g. wireless router connected to the Internet via fiber), and corporate networks (e.g. Ethernet). They employ different physical media, either wired or wireless.

Both wired and wireless technologies have remarkably developed toward higher data rates, also improving the quality of service in terms of latency and reliability. The currently main wired technologies are twisted pair copper wire, coaxial cables, and fiber optics. The main wireless technologies employ electromagnetic waves to build radio channels allocated in the frequency (spectrum) domain. The radio channels are usually terrestrial as in cellular networks, but there are also satellite communication links including geostationary and low-earth orbiting (LEO) satellites; underwater wireless communications, in turn, are not as widespread as the other two approaches.

The network core can be simply defined as [3] *(…) the mesh of packet switches and links that interconnects the Internet's end systems.* In other words, the core network function is to physically enable the logical connections between end systems and their applications. As indicated in Figure 9.5, the main elements of the network core are packet switches, namely link layer switches and routers. The function of those devices is to transfer digital data from one point of the core network to another aiming at the final destination in a decentralized manner (i.e. the end system that the message is addresses), but considering that there is no central controller to determine the optimal route (which would be nevertheless unfeasible because of the computational complexity of such a task; see Chapter 6).

The strength of the layered approach and the message encapsulation can be fully appreciated here. At the source end system, the considered application generates a *message* that needs to be transferred to a given destination end system. The message generated by the application layer is divided into smaller data chunks called *segments*, which are then combined with a transport layer header, which provides the instructions of how the original data shall be reassembled at the destination.

This layer function is to construct a logical link between end systems in order to transport application layer messages from a source to a destination. TCP and User Datagram Protocol (UDP) are the Internet transport layer protocols. This step is still related to the network edge, and the data segments still require additional instructions to reach their destination through the network core.

The following steps of the data encapsulation process solve this issue. At the network layer, a new header is added to the data segments containing the details about the route to reach the final destination, forming then *datagrams* or *data packets*; the network layer protocol of the Internet is the IP. This layer is divided into a control plane related to the "global level" routes and the overall network topology (including the information needed to build routing tables) and a data plane related to forwarding from one node to another based on routing tables following the control plane policies. These two planes have traditionally been integrated into the physical router device. Currently, the trend is to consider these planes independently through *software-defined networking* (SDN).

To accomplish the transmission of data packets from one node (either a host or a router) to the next one considering that there are several nodes between the source and the destination, additional specification is needed related to the specific physical communication link. A new header is added to the datagram at the link layer, then forming a *frame*. There are several link layer protocols like Ethernet and IEEE 802.11. Medium access protocols, error detection, and error correction are related to the link layer. The physical layer is related to how the data frame is coded to be physically transmitted as modulated digital data.

This is an extremely brief introduction to data networks, specifically to the Internet, which indicates the fundamentals of how such a system works. For interested readers, the textbook [3] offers a pedagogic introduction to the field. In the following subsection, we will turn our attention to wireless communications, specifically the novelties brought by the 5G with respect to machine-type of traffic generated by the massive and still increasing adoption of devices connected in the Internet.

9.2.3 IoT, Machine-Type Communications, and 5G

Wireless communications have been successfully deployed for some time and considering different end applications including localization services with satellite communications, public local Internet access with WiFi, and mobile networks for voice communication covering large areas. This last example refers to cellular systems, whose historical development is punctuated by different generations. Figure 9.6 illustrates the key differences of the existing five generations, from the first generation (1G) to the 5G that is currently being deployed worldwide.

The 1G was based on analog modulation for voice communication; it was developed to provide mobile telephone service only. The 2G implemented digital

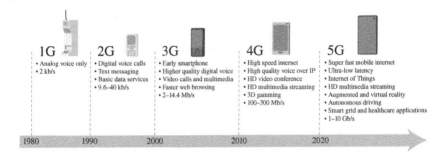

Figure 9.6 Current five generations of cellular systems in 2021.

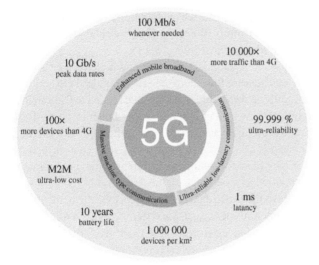

Figure 9.7 5G applications and their requirements in terms of rate, reliability, and latency.

transmission techniques, also allowing Short Message Service (SMS) and relatively limited Internet connectivity. The 3G provided further improvements to guarantee Internet connectivity for mobile devices and also included services of video calling and Multimedia Messaging Service (MMS). The main focus of the 4G is on the (remarkably) improved quality of the Internet access, even allowing video stream of full high definition (full HD) video. The current 5G deployment targets not only human-type communications and related applications, but also MTCs arising from the massive connectivity of devices and specific critical applications.

Figure 9.7 depicts the key 5G applications, which are mapped in three axes referring to their demand of throughput, number of communicating devices, and required performance in terms of reliability and latency. There is then a large set of applications that the 5G shall offer as a standardized wireless solution, including the coexistence of human users with their download (downlink) dominated traffic composed of relatively large messages (e.g. video stream) and machines with their upload (uplink) dominated traffic composed of short, usually periodic, timestamped messages (e.g. temperature sensors). MTCs usually refer to two extreme types of applications, namely (i) massive machine-type communication (mMTC), where potentially millions of sensors need to transmit their data to be stored and further processed in large data centers and supercomputers, and (ii) ultrareliable low latency communications (URLLC) to be used in critical applications that are related to, for example, feedback control loops in industrial automation and self-driving cars.

The mMTC are frequently associated with cloud computing, where the data from the machine-type devices are first sent to cloud servers that store and process such *big data* that become available to be retrieved by potentially different applications. For example, temperature sensors that send the information to the cloud server of a meteorological research center, whose data can be used either by humans to verify what would be a suitable dress for the day or by a supercomputer to simulate the whether conditions.

URLLC, in turn, involve applications requiring a latency lower than 1 ms and a reliability of 99.9999%; there are also other applications with very strict latency and reliability constraints but which are not considered part of this class; for example, a communication link for the feedback control loop of a robot arm. In any case, applications with such requirements are unfeasible to be deployed by using cloud computing because of the delays related to the network core. In this case, edge computing offers a suitable alternative. Figure 9.8 illustrates the difference between these two paradigms.

There are nowadays, in 2021, huge research and development efforts to deploy 5G, considering both the radio access technologies and radio access networks (both at the network edge), as well as aspects related to the network core via software-defined networks. A comprehensive introduction of wireless systems is provided in [4]. In the following section, we will dive into special techniques developed to process the large amount of data acquired by sensors and other devices, which today are broadly known as AI and machine learning.

9.3 Artificial Intelligence and Machine Learning

The term AI has a long history, which might be even linked with ancient myths. As the name indicates, AI should be contrasted with natural intelligence, particularly human intelligence [10]. Of course, this topic has a philosophical appeal

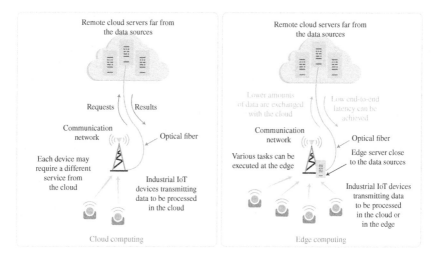

Figure 9.8 Cloud vs. edge computing. Source: Adapted from [9].

associated with the classical questions of idealist philosophy about Freedom, Autonomy, Free Will, Liberty, and Meaning of Life. Although such metaphysical questions are still quite relevant, AI has also acquired a more scientific status closely related to the cybernetic ideas that emerged after the World War II, although the main proponent of AI as an independent research field, John MaCarthy, decided to use this name to avoid the association with Wiener's cybernetics; a brief personal note indicates the climate of the discussions [11].

As indicated in Chapter 1, this book tries to avoid those problems by establishing the object of the proposed theory. In our understanding, the term AI is misleading and is actually based on a naive, mostly behaviorist view of what *intelligence* is, which was also affected by a reductionist epistemology associated with methodological individualism [12]. In this book, what one would call AI systems, or AI-based systems, should be associated with self-developing reflexive–active systems as presented in Chapter 7. Besides, AI is generally related to agents (artificial entities that possess both intelligence to manipulate data and ability to act and react), agent-based models, and multiagent systems that are based on autonomous elements that sense the environment, process these data, and react. This topic is particularly covered in Chapter 6, where decision-making and acting processes are presented.

Machine learning (ML), as a term, also has similar issues to the ones we just presented and is usually classified as a branch of AI. However, ML has a more technical flavor supported by established disciplines, such as probability, statistics, optimization, linear algebra, dynamical systems, and computer sciences. In my view, a more correct name for ML would be either data processing theory or applied data processing, although I also acknowledge that the use of analogies from other disciplines may help the technological development. Nevertheless, the aim is not to change how this field is widely known, and this book will retain the usual terminology. These first paragraphs should be read as a warning note about potential misinterpretations or transpositions that these names might lead to. In the following, the basics of ML will be presented employing the approach taken by Jung [13].

9.3.1 Machine Learning: Data, Model, and Loss Function

In [13], Alex Jung defines the three components of ML, namely data, model, and loss. **Data** refer to attributes of some system that can be quantified in some way that are related to **features (inputs)** and **labels (outputs)**. **Models** refer to (computationally feasible) mathematical functions that map the feature space into a label space; models are also referred to as **hypothesis space** and the functions are known as **predictors** or **classifiers**. **Loss function** refers to the function that is employed to quantify how well a predictor or a classifier works in relation to the actual outputs. In the following, we will propose a formal example of a ML problem statement with a solution.

9.3.2 Formalizing and Solving a ML Problem

Any ML problem involves prediction or classification about a given system or process based on quantified attributes that are potentially informative data (see Chapter 4). These data are then split into two groups, namely inputs (features) and outputs (labels) of the ML problem. Before the prediction or classification, there is a *training* phase where, for example, known input–output pairs serve as the way to find optimal model in terms of a given loss function. In the following, we will formalize a simple ML problem.

Definition 9.2 *Linear regression as a ML problem.* Let $x[k] \in \mathbb{R}$ and $y[k] \in \mathbb{R}$ be two quantifiable attributes of a given system Φ, both indexed by a parameter $k \in \mathbb{N}$. The ML problem is to provide predictions $\hat{y}[k] = h(x[k])$ where $h : \mathbb{R} \to \mathbb{R}$ represents the model used to map the $x[k]$ to $y[k]$. If we assume that the relation between $x[k]$ and $y[k]$ is linear so that $\hat{y}[k] = h(x[k]) = w_1 x[k] + w_0$, where $w_1, w_0 \in \mathbb{R}$ are the weights used to define the impact of $x[k]$ in $y[k]$ and the offset value (i.e. the value of $y[k]$ when $x[k] = 0$). Consider that there are m available pairs of $(x[k], y[k])$ that we used to find w_1 and w_0 that minimize the loss function, which is here defined as the mean square error (MSE) $(1/m) \sum_{i=1}^{m} (y[i] - h(x[i]))^2$. The

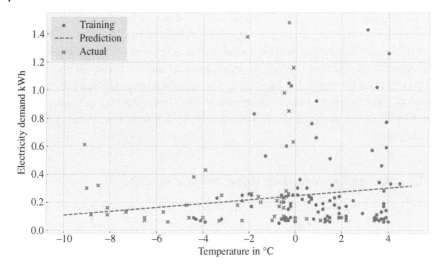

Figure 9.9 Predicting the electricity demand from air temperature based on linear regression.

quality of the predictor $\hat{y}[k]$ can be tested whenever the true label $y[k]$ is available by using the square error or by the MSE if more true labels are available.

Example 9.2 *Air temperature and electricity consumption.* Consider two timestamped datasets: (a) the average temperature during a given hour and (b) the average electric power consumption in a household in a given hour. We employ Definition 9.2 as follows.

- **Data:** Dataset (a) is the feature denoted as $x[k]$ and dataset (b) is the label denoted as $y[k]$.
- **Model:** $\hat{y}[k] = h(x[k]) = w_1 x[k] + w_0$
- **ML problem:** Find w_1 and w_0 that minimize the MSE loss function $(1/m_t) \sum_{i=1}^{m_t} (y[i] - h(x[i]))^2$ in a training dataset with m_t elements $(x[k], y[k])$.
- **Precision of the learned predictor:** Evaluate the quality of the predictions using the MSE loss function $(1/m_q) \sum_{i=1}^{m_q} (y[i] - h(x[i]))^2$ in a quality testing dataset with m_q elements $(x[k], y[k])$, which are different from the other m_t elements.

Figure 9.9 depicts the results using the air temperature and electricity demand data in a winter month in Finland. The linear prediction (dashed light grey line) was constructed based on training data (circles), resulting in $\hat{y}[k] = 0.0137 \, x[k] + 0.245$. The quality of the prediction can be visualized by actual data (Xs) in the same winter month.

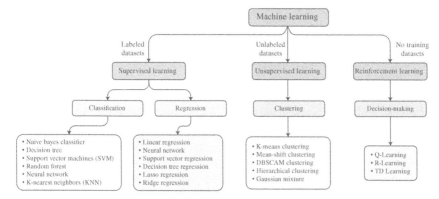

Figure 9.10 ML methods classified as a combination of data, model, and loss that are employed to solve different types of problem.

The minimization of MSE in linear equations is well established and easy to implement in software. Low dimensional datasets, linear models, and availability of labels are not the rule in ML problems, though. As mentioned in [13], there are several ML methods, which are defined by the specific combination of data, model, and loss, that better suit the system and process in hand. We will briefly discuss them next.

9.3.3 ML Methods

There are several ML methods that are employed to solve specific classification and prediction problems related to a PS or process. Such methods can be specified by a particular combination of data, model, and loss that are used to solve the ML problem. Figure 9.10 illustrates the classes of learning based on (i) the type of data available from training purposes and (ii) the type of the ML problem to be solved.

In terms of data, ML methods can be classified as supervised or unsupervised learning, the first one being related to the availability (or existence) of data about the label variables. The loss function can be based on squared error, logistic loss, or 0/1 loss. The model can be linear maps, piecewise constant, or neural networks. Each combination leads to a particular ML method, which, in turn, is suitable to perform a given computational task. For example, the case presented in the previous subsection is a supervised ML that considers a linear map model with a squared error loss function, defining a *linear regression* method to predict the particular value of a label for a given value of the feature.

It is also important to mention that there is another class of learning different from supervised and unsupervised ML, which is known as *reinforcement learning*. This method considers that a given agent needs to accomplish a predetermined goal in a trial-and-error fashion by indirectly learning through its interactions with the environment. The success and failure of each action in terms of the goal to be achieved is defined by a utility (or loss) function, which quantifies the level of success of each action tried by the agent. Interestingly, each action is determined by the particular model used by the agent. Reinforcement learning is then considered a third class of ML, where the model is dynamically constructed in a trial-and-error fashion by (indirectly) quantifying the success (failure) of an action in terms of the utility (or loss) function. For each new input data (feature), reinforcement learning methods first produce an estimated optimal model that is used to predict the respective output and the respective action. Note that the loss function can only be evaluated considering the actually selected model.

Figure 9.10 provides a good overview of ML methods and their classification, also indicating the names of some well-known techniques. For more details, there is plenty of literature on ML methods as introduced in [13] and references therein. In the following section, we will cover another important fundamental aspect of CPS, which also involves questions related to data networks and ML methods; this refers to decentralized computing paradigms and decentralized data storage.

9.4 Decentralized Computing and Distributed Ledger Technology

While the physical and logical topology of the Internet is decentralized, the majority of applications work in a more centralized manner where more capable (hardware) elements are employed to store a massive amount of data of a huge number of individual users or to perform computationally complex calculations. This is the idea behind the concept of cloud computing. As we have mentioned in Section 9.2, other computing paradigms are emerging where elements at the network edge are employed to perform computations or to build a trustworthy database without the need of a third party that certifies transactions mediated by data. Two specific techniques – federated learning and blockchain – will be briefly described because they illustrate the typical advantages and drawbacks of decentralization. Note that these approaches are different from local storage and computation that are carried out by individual machines, such as personal computers, mobile phones, and powerful desktops. In contrast, they are designed considering a network of interacting elements that jointly constitute a specific system that performs a peculiar function.

9.4.1 Federated Learning and Decentralized Machine Learning

Consider the following scenario: several IoT devices that measure different attributes of a given system, being then connected to the Internet. This large

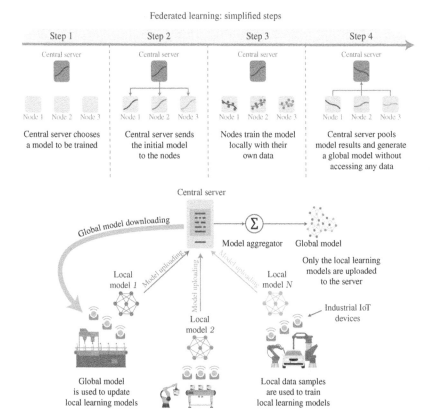

Figure 9.11 Federated learning.

amount of data shall be processed to train a supervised ML, whose result should be delivered back to the different IoT devices that will provide predictions to other applications. Figure 9.8 provides two possible solutions for this problem: (i) cloud computing where the ML model is trained in a supercomputer somewhere at the cloud server, or (ii) edge computing where the ML model is trained in an edge server closer to the data source (e.g. co-located with a base station). In both cases, the end systems do not actually cooperate; they just provide the data to be processed and then receive the ML model to be used.

Federated learning is different: a central server (either a cloud or an edge server) is used to organize the end systems to perform computations to define local models, which are then used by the central server to generate a global model to be used by the end users. Figure 9.11 depicts an example of the federated learning method. The main advantages of federated learning are [9]: less use of communication resources because less data need to be transmitted, lower communication delays if the associated central server is located at the network edge, and higher privacy guarantees because only local models are sent to the central server (not the raw data). The main drawbacks are related to the relatively small computational power of end users compared with the cloud or edge servers, and the energy required to perform computations that might be prohibitive.

The star topology employed by the federated learning may also be problematic, mainly considering a scenario where the end systems are IoT devices wirelessly connected to the data network via radio. In this scenario, other questions related to the link reliability and the coverage radio become important. A possible solution to this issue is collaborative federated learning, where different network topologies would be allowed [14]. This case would even include a more extreme case of a fully distributed, peer-to-peer topology.

In addition to federated learning, there are also several other methods to perform distributed machine learning, as surveyed in [15]. In particular, the most suitable choice between centralized, decentralized, or even fully distributed ML implementations depends on different aspects of the specific application and its performance requirements in terms of, for instance, accuracy and convergence time, and also of end user limitations in terms of, for instance, computational power, data storage, availability of energy, and data privacy.

9.4.2 Blockchain and Distributed Ledger Technology

The blockchain technology was first developed to guarantee the trustworthiness of the cryptocurrency *bitcoin* without the need of a third party. The technical solution was described in 2008 as follows [16]:

> What is needed is an electronic payment system based on cryptographic proof instead of trust, allowing any two willing parties to transact directly with each other without the need for a trusted third party. Transactions that are computationally impractical to reverse would protect sellers from fraud, and routine escrow mechanisms could easily be implemented to protect buyers. In this paper, we propose a solution to the double-spending problem using a peer-to-peer distributed timestamp server to generate computational proof of the chronological order of transactions. The system is secure as long as honest nodes collectively control more CPU power than any cooperating group of attacker nodes.

In this case, the function of the blockchain is to build, in a distributed way, a distributed ledger where the transaction data have their integrity verified without a central element. In a much less technical (and heavily apologetic) comment article [17], the author explains the reasons why blockchain and bitcoin are appealing:

> Bitcoin's strength lies in how it approaches trust. Instead of checking the trustworthiness of each party, the system assumes that everyone behaves selfishly. No matter how greedily traders act, the blockchain retains integrity and can be trusted even if the parties cannot. Bitcoin demonstrates that banks and governments are unnecessary to ensure a financial system's reliability, security and auditability.

A more detailed discussion about the pitfalls of such a way of thinking will be provided in Chapters 10 and 11, where different applications and aspects beyond technology will be discussed. What is important to mention here is that, while distributed learning explicitly involves a certain level of cooperation between computing entities, blockchain is clear about its generalized assumption that all elements are selfish, competitive, and untrustworthy. Peers do not trust each other, neither a possible third party that would verify transactions between peers; they rather trust in the system effectiveness to perform the verification task in a distributed manner.

In technical terms, blockchain is designed to securely record transactions (or data in general) in a distributed manner by chaining blocks. Blocks corresponds to verified transactions (or data). To be accepted, a new block needs to be verified by the majority of the peers in the system based on a cryptographic hash function. Upon its acceptance, the new block is uniquely identified by a hash that links it to the previous block. In this way, the more blocks there are in the chain, the more secure is the distributed ledger in relation to tampering data. Details of the implementation can be found in [16].

Distributed ledger technology (DLT) usually refers to a generalization of blockchain (although some authors consider DLT and blockchain as synonyms). In any case, Figure 9.12 illustrates the main principles of DLT and blockchains as described above but without being restricted to its original design. The references [18] and [19] present interesting reviews of DLTs and their different design options and promising application cases beyond cryptocurrencies like bitcoin. Examples of applications are smart contracts, supply chain product tracking, and health care data.

Another important DLT application comes from IoT. The reference [20] revisits the main DLTs that are used for IoT deployments. For CPSs, DLT might be useful to guarantee the trustworthiness of sensor measurements, but this brings new challenges for the data network design because the verification process increases the network traffic, mainly at the edge with remarkable changes in the downlink [21]. As in the case of decentralized ML, the most suitable DLT design depends on tradeoffs related to, for instance, computation capabilities,

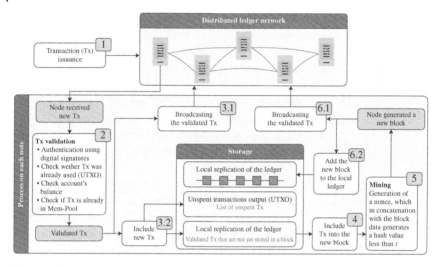

Figure 9.12 Distributed ledger technology. Source: Adapted from [19].

issues related to data transmission and energy consumption, and individual data storage. Nevertheless, the most important task before any deployment is to actually identify whether a specific application indeed requires a DLT solution considering its main presupposition: peers do not trust each other while fully trusting the technical implementation.

9.5 Future Technologies: A Look at the Unknown Future

Despite the ongoing discussions of today (2021) about what the 6G will be, which (new) applications shall be covered by the standard, and what is actually technically feasible to deploy, this section covers two paradigms that, in my view, would bring more fundamental changes in the technical domain. The first one is the Quantum Internet and the second is the IoBNT. This section ends with a brief discussion about what is coming with Moore's law.

9.5.1 Quantum Internet

Quantum computing is already a reality where computers are designed to use phenomena from quantum physics (e.g. superposition and entanglement) to process data mapped as *qubits*, where the states "0" and "1" of classical digital systems can be in a superposed state and be teletransported. In 2021, this technology is not yet widespread and only a few institutions have the (financial and technical) capabilities to have them. The main advantage of quantum computing is to solve at much higher speeds problems that are hard for classical computers, particularly combinatorial problems.

The Quantum Internet, in turn, does not need to be, in principle, an Internet composed of quantum computers. In a paper called *Quantum internet: A vision for the road ahead* [22], the authors explain the concept as follows.

> The vision of a quantum internet is to fundamentally enhance internet technology by enabling quantum communication between any two points on Earth. Such a quantum internet may operate in parallel to the internet that we have today and connect quantum processors in order to achieve capabilities that are provably impossible by using only classical means.

At this stage, the authors already foresee some important applications, such as secure communications using quantum key distribution (QKD) to encrypt the message exchanges and extremely precise time synchronization. Figure 9.13 illustrates the proposed six developmental stages to construct a full-fledged Quantum Internet. An important, sometimes not obvious, remark is that new applications, which are impossible to predict today, will surely appear throughout the development of quantum technologies. Because such large-scale deployment is both uncertain and highly disruptive, it deserves to be mentioned here: the deployment of the Quantum Internet would have an unforeseeable impact on the data and decision layers of CPSs. An interested reader could also refer to [23] for a more detailed survey of such a new technology.

9.5.2 Internet of Bio-Nano Things

In Chapter 3, the concept of biological information was briefly introduced to indicate the existence of (bio)chemical signals that constitute living organisms. The IoBNT has been proposed in [24] as a term to indicate how cells, which are the substrates of the Bio-Nano Things, could be thought to have similar functionalities to IoT devices, and thus, explicit interventions could be designed accordingly. The Bio-Nano Things would have the following functional units: controller, memory, processor, power supply, transceivers, sensors, and actuators; all mapped to specific molecules and molecular processes at the cell. The main idea of the IoBNT is to engineer molecular communications to modify the biological processes in a desired manner, which might support the development of new ways to deliver medical treatments.

Moreover, the authors of [24] aim at scenarios where the bio-nano things are connected not only among themselves but also on the Internet. This idea would be

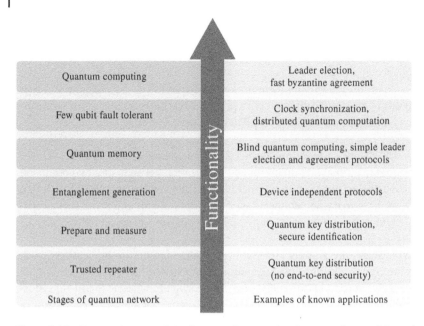

Figure 9.13 Proposed stages of the Quantum Internet development. Source: Adapted from [22].

a further extension of the already established man–machine interfaces (including brain–machine interfaces) and wearables of all kinds (e.g. smart watches and patches). What makes the IoBNT both promising and scary is the level of intrusiveness of the technology. Explicitly control molecular communications in a living body to allow coordination between cells, potentially through the Internet if suitable interfaces are developed, open up a series of possible new treatments for diseases and medical practices. On the other hand, the social impact can be fairly high; corporations and governmental institutions (including military and security agencies) will certainly find new ways to make profit and build even more direct and individualized technical solutions to analyze, profile, and control persons using their biological data. In this case, the development of the IoBNT is uncertain for technical and, above all, social and ethical reasons; nevertheless, if eventually deployed and widespread, such a technology would allow a large scale of cyber-physical biophysical systems interacting with other CPSs.

9.5.3 After Moore's Law

Gordon P. Moore – engineer, businessman and cofounder of one of the biggest semiconductor manufactures in the world – stated in 1975 an empirical characterization about the density of microchips pointing out that the density of

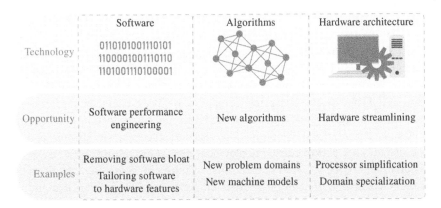

Figure 9.14 Performance gains after Moore's law. Source: Adapted from [26].

transistors in integrated circuits will approximately double every two years [25]. This statement worked as a sort of self-fulfilling prophecy or benchmark that is known as Moore's law. The miniaturization of computers is usually seen as a consequence of Moore's law. After several decades guided by such a forecast, the density of integrated circuits is now reaching its physical limits, and thus, setting an end to Moore's law. In other worlds, it seems that the gains experienced in the last fifty years, when a 2020 mobile phone is probably more powerful than a 1970 mainframe supercomputer, are over. The question that remains is: how can the huge performance gains that were predicted by Moore's law be realized in the future to come?

The reference [26] provides a well-grounded view of how to keep improving the efficiency of computers in such a new technical world. The authors' main argument is very well summarized by the title: *There's plenty of room at the Top*. The main idea is that, while gains at the bottom (densification of integrated circuits) is halting, many improvements can be made in software performance, algorithm design, and hardware architecture.

Figure 9.14 illustrates their view. These indications are very well grounded, also aligned with some comments presented in [15] about how specialized hardware architectures can improve the performance of distributed ML taken lessons from the parallelization of computing tasks studied by high performance computing research communities. The authors also identified potential new paradigms at the bottom, such as quantum computing, graphene chips, and neuromorphic computing, which might also be seen as potential enablers of the Quantum Internet and IoBNT. In summary, the years to come will determine what the (unwritten) long-term future of computing (and CPSs) will be.

9.6 Summary

This chapter covered two fundamental enabling technologies for CPSs, namely data networks and AI. The focus was specifically given to the Internet-layered architecture that can be seen as the way to physically establish logical links to transfer data and machine learning techniques to process data to be further used for prediction, classification, or decision. A very brief introduction to emerging topics like decentralized computing and distributed ledgers was also presented, because they are part of current deployment discussions of different CPSs. This chapter ended with a look at the still unclear future by discussing two different nascent technologies, Quantum Internet and IoBNT) and the future of computing after Moore's law. It is worth saying that our exposition was extremely short and oversimplified, and therefore, interested readers are invited to follow the references cited here to dive deeper into the different specialized technological domains.

Exercises

9.1 Random access for IoT. Most IoT devices are connected to the Internet through wireless links, which are defined by slices in the electromagnetic spectrum defined by frequencies with a given bandwidth. The use of the same frequency by two or more devices may cause destructive interference that leads to collisions and communication errors. The potentially massive number of IoT devices may lead to a situation where both centralized resource allocation is unfeasible and medium access without any contention/coordination has an overly high level of interference, which makes a successful transmission impossible.

One solution consists of using random access medium access control (MAC) protocols. There are four candidates:

- **Pure ALOHA:** Whenever a packet arrives at the node, it transmits.
- **Slotted ALOHA:** The time is slotted, and whenever a packet arrives, it is transmitted in the following time slot.
- **Nonpersistent Carrier Sensing Multiple Access (CSMA):** If a packet arrives, the node senses the channel; if no transmission is sensed, then the transmission starts; if a transmission is detected, the transmission is backed off for a random time when the sensing process takes place again. This process is repeated until the packet is transmitted.
- **1-persistent CSMA:** If a packet arrives, the node senses the channel; if no transmission is sensed, then the transmission starts; if a transmission

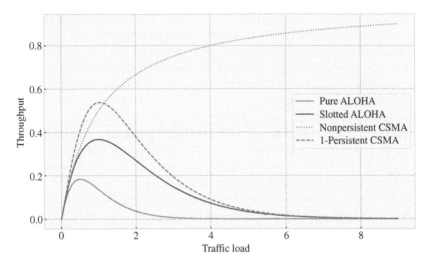

Figure 9.15 Throughput (from 0 to 1) as a function of traffic (expected packets per transmission time) for ALOHA and CSMA MAC protocols. The details of the model and equations are found in [27].

is detected, the node keeps sensing until the channel becomes free when the packet is transmitted.

Figure 9.15 compares the performance of these four protocols in terms of throughput evaluated as the probability that a transmission is successful as a function of network traffic defined as the expected number of packets generated during the transmission time, which is considered fixed and related to data packets of the same size.

(a) Explain the results presented in Figure 9.15 based on the protocol description.

(b) CSMA outperforms ALOHA in terms of throughput, but this metric cannot capture the impact of these MAC protocols in the expected delay. Provide a qualitative comparison between CSMA and ALOHA in terms of expected delay.

(c) The nonpersistent and the 1-persistent CSMA have a clear performance gap in terms of throughput, but also in terms of expected delay because in the first case every time that the channel is detected busy the transmission must wait a random time while the second case the node keeps sensing the channel until is free, but collisions may happen when more than one node is waiting to transmit. Propose a solution based on randomization that shall control the performance trade-off between these two classes of CSMA. *Hint:* The solution is known as *p*-persistent CSMA.

(d) Although this analysis is theoretical, those MAC protocols are indeed implemented in actual wireless networks. The master's thesis [28] provides a good overview of wireless technologies for IoT communications followed by an example of a feasibility study of different technologies for transmitting data of weather station sensors. The task is to read this more practical research work.

9.2 Designing a sensor network to measure air pollution. The government of a given city decided to distribute several air pollution sensors in a specific industrial region. These sensors have their energy supplied by a battery and can perform computations and store data, although their capabilities are quite limited. They are also wirelessly connected to the Internet via a 4G wireless network. Assess qualitatively the benefits and drawbacks of the following possible design solutions.

(a) Data processing architecture (centralized or decentralized) to construct a dynamic ML model to predict the hourly levels of pollution for the upcoming 12 hours based on the past data.

(b) Type of database to record new measurements considering the options of (i) one individual data storage server, (ii) a DLT where the data are stored in several desktop machines, or (iii) a DLT where the data are stored in the sensors. The server, the desktops, and the sensors are all owned by the city.

(c) Think about the conditions and requirements that make a centralized ML solution with DLT in desktop machines the most suitable solution.

9.3 Imagine the future. This task is a creative one. Write a two- to five-page science fiction narrative about the future focusing on the technologies described in Section 9.5. The text can be an utopia, a dystopia, or a more realistic one.

References

1 Aglietta M. A Theory of Capitalist Regulation: The US Experience. vol. 28. Verso; 2000.

2 Gerovitch S. From Newspeak to Cyberspeak: A History of Soviet Cybernetics. MIT Press; 2004.

3 Kurose JF, Ross KW. Computer Networking: A Top-Down Approach. Pearson; 2016.

4 Popovski P. Wireless Connectivity: An Intuitive and Fundamental Guide. John Wiley & Sons; 2020.

5 Madhow U. Introduction to Communication Systems. Cambridge University Press; 2014.

6 Huurdeman AA. The Worldwide History of Telecommunications. John Wiley & Sons; 2003.

7 Osseiran A, Monserrat JF, Marsch P. 5G Mobile and Wireless Communications Technology. Cambridge University Press; 2016.

8 Dang S, Amin O, Shihada B, Alouini MS. What should 6G be? Nature Electronics. 2020;3(1):20–29.

9 Narayanan A, De Sena AS, Gutierrez-Rojas D, Melgarejo DC, Hussain HM, Ullah M, et al. Key advances in pervasive edge computing for industrial Internet of Things in 5G and beyond. IEEE Access. 2020;8:206734–206754.

10 Bringsjord S, Govindarajulu NS, Zalta EN, editor. Artificial Intelligence. Metaphysics Research Lab, Stanford University; 2020. https://plato.stanford.edu/archives/sum2020/entries/artificial-intelligence/.

11 McCarthy J. Review of the question of artificial intelligence. Annals of the History of Computing. 1988;10(3):224–229.

12 Heath J, Zalta EN, editor. Methodological Individualism. Metaphysics Research Lab, Stanford University; 2020. https://plato.stanford.edu/archives/sum2020/entries/methodological-individualism/.

13 Jung A. Machine Learning: The Basics. Springer; 2022. https://link.springer.com/book/10.1007/978-981-16-8193-6.

14 Chen M, Poor HV, Saad W, Cui S. Wireless communications for collaborative federated learning. IEEE Communications Magazine. 2020;58(12):48–54.

15 Verbraeken J, Wolting M, Katzy J, Kloppenburg J, Verbelen T, Rellermeyer JS. A survey on distributed machine learning. ACM Computing Surveys (CSUR). 2020;53(2):1–33.

16 Nakamoto S. Bitcoin: a peer-to-peer electronic cash system. Decentralized Business Review. 2008:21260. https://bitcoin.org/bitcoin.pdf.

17 Chapron G. The environment needs cryptogovernance. Nature News. 2017;545(7655):403.

18 Liu X, Farahani B, Firouzi F. Distributed ledger technology. In: Intelligent Internet of Things. Firouzi F, Chakrabarty K, and Nassif S, editor, Springer; 2020. p. 393–431. https://link.springer.com/chapter/10.1007/978-3-030-30367-9_8.

19 Sunyaev A. Distributed ledger technology. In: Internet Computing. Springer; 2020. p. 265–299. https://link.springer.com/chapter/10.1007/978-3-030-34957-8_9.

20 Wang X, Zha X, Ni W, Liu RP, Guo YJ, Niu X, et al. Survey on blockchain for Internet of Things. Computer Communications. 2019;136:10–29.

21 Nguyen LD, Kalor AE, Leyva-Mayorga I, Popovski P. Trusted wireless monitoring based on distributed ledgers over NB-IoT connectivity. IEEE Communications Magazine. 2020;58(6):77–83.

22 Wehner S, Elkouss D, Hanson R. Quantum internet: a vision for the road ahead. Science. 2018;362(6412).

23 Singh A, Dev K, Siljak H, Joshi HD, Magarini M. Quantum internet-applications, functionalities, enabling technologies, challenges, and research directions. IEEE Communication Surveys and Tutorials. 2021;23(4):2218–2247. https://doi.org/10.1109/COMST.2021.3109944.

24 Akyildiz IF, Pierobon M, Balasubramaniam S, Koucheryavy Y. The internet of bio-nano things. IEEE Communications Magazine. 2015;53(3):32–40.

25 Moore GE. Progress in digital integrated electronics [Technical literature, Copyright 1975 IEEE. Reprinted, with permission. Technical Digest. International Electron Devices Meeting, IEEE, 1975, pp. 11–13.]. IEEE Solid-State Circuits Society Newsletter. 2006;11(3):36–37.

26 Leiserson CE, Thompson NC, Emer JS, Kuszmaul BC, Lampson BW, Sanchez D, et al. There's plenty of room at the top: what will drive computer performance after Moore's law? Science. 2020;368(6495):eaam9744. https://doi.org/10.1126/science.aam9744.

27 Kleinrock L, Tobagi F. Packet switching in radio channels: Part I-carrier sense multiple-access modes and their throughput-delay characteristics. IEEE Transactions on Communications. 1975;23(12):1400–1416.

28 Mäki V. Feasibility evaluation of LPWAN technologies: case study for a weather station; 2021. M.Sc. thesis. Lappeenranta–Lahti University of Technology. Available at: https://lutpub.lut.fi/handle/10024/162347.

10

Applications

This chapter addresses different cyber-physical systems (CPSs) enabled by the information and communication technologies (ICTs) presented in the previous chapters. The idea is to employ the concepts presented so far in order to analyze several examples of already existing applications related to industrial plants, residential energy management, surveillance, and transportation. The focus here will be on the CPS itself without considering aspects that are beyond the technology, such as governance models and social impacts, which will be discussed in Chapter 11. Specifically, the following applications will be covered: (i) fault detection in the Tennessee Eastman Process (TEP) [1], (ii) coordination of actions in demand-side management actions in electricity grids [2], (iii) contention of epidemics [3], and (iv) driving support mobile applications [4]. The first two will be dealt with in more detail, while the other two will be part of exercises.

10.1   Introduction

Back in Chapter 1, it was argued that the deployment of CPSs does not require a theory specially constructed to apprehend their particularities. After a long tour covering basic concepts (Chapters 2–6), the definition of the three layers of CPSs (Chapters 7 and 8) and key enabling ICTs (Chapter 9), actual real-world realizations of CPSs related to specific applications will be presented here. The approach taken considers a generalization of the framework introduced in [1, 5], where seven questions are employed to determine the design aspects of a given CPS considering its peculiar function articulated with specific technical interventions that might be considered. Here we add one more question and reformulate the others.

Figure 10.1 depicts the three-layer model considering the eight guiding questions presented in Table 10.1. The system demarcation as shown in Chapters 2 and 7 together with such questions aims at defining the specific characterization of the CPS under consideration as a step to build or understand an actually

Cyber-physical Systems: Theory, Methodology, and Applications, First Edition. Pedro H. J. Nardelli.

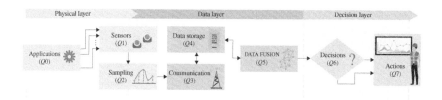

Figure 10.1 Illustration of the three layers of CPSs and their physical deployment.

Table 10.1 Guiding questions to study and/or design CPSs.

Q#	Topic	Related question
Q0	Applications	What are the applications to be implemented and their relation to the CPS function?
Q1	Sensors	What kinds of sensors will be used? How many of each can be used? Where can they be located?
Q2	Samples	Which type of sampling (data acquisition) method will be used? How data are coded?
Q3	Communication	Which type of communication system (access, edge and core technologies) will be used?
Q4	Data storage	Where are the data from sensors stored and processed (locally, in the edge and/or in the cloud)?
Q5	Data fusion	Will the data be clustered, aggregated, structured, or suppressed? How?
Q6	Decisions	How will the (informative) data be used to make decisions?
Q7	Actions	Are there actuators and/or human interfaces? If yes, how many and where?

operating CPS with its reflexive–active self-developing dynamics. Note that the CPS demarcation based on its peculiar function and the answers to the proposed questions are highly related, indicating, for instance, feasible design options and potential improvements.

In this case, Table 10.1 is constructed by acknowledging the fact that a given CPS may support, in addition to its peculiar functions, other applications that are designed either to guarantee such a functionality or to improve its effectiveness based on measurable attributes and possible interventions. As to be discussed in the following sections, the TEP can be considered a CPS where a dedicated application to detect faults can run, and thus, actions by the responsible personnel can be taken accordingly [1, 6]. Distributed coordination of actions in demand-side management where individual devices react to a perturbation in a physical system can also be designed as a CPS [7]; this approach of a cyber-physical energy system will be further extended in Chapter 11 when presenting the Energy Internet [7, 8] where part of the electricity interchanges in the grid are to be virtualized as energy packets to be governed as a commons [9].

10.2 Cyber-Physical Industrial System

In this section, the focus is on CPSs that are deployed in industrial plants in order to improve their operation. The cyber-physical industrial system addressed here is the TEP benchmark widely used in the literature of process control engineering, where faults need to be detected and classified based on acquired data. The details of the TEP are presented next.

10.2.1 Tennessee Eastman Process

The TEP was initially proposed by researchers of the Eastman Chemical Company [6]. The idea was to provide a realistic simulation model to reflect typical challenges of process control considering nonlinear relations. The physical process is characterized by the following set of irreversible and exothermic chemical reactions:

$$A_{(g)} + C_{(g)} + D_{(g)} \rightarrow G_{(liq)}, \qquad \text{Product 1},$$
$$A_{(g)} + C_{(g)} + E_{(g)} \rightarrow H_{(liq)}, \qquad \text{Product 2},$$
$$A_{(g)} + E_{(g)} \rightarrow F_{(liq)}, \qquad \text{Byproduct},$$
$$3D_{(g)} \rightarrow 2F_{(liq)}, \qquad \text{Byproduct},$$

where the process has four reactants (A, C, D, E), two products (G, H), one byproduct (F), and one inert (B), resulting in eight components. The process takes place in five units, namely reactor, condenser, vapor–liquid separator, recycling compressor, and stripper. Figure 10.2 presents a schematic depiction of the TEP.

The details about the process can be found in [6]. What is important here is that the proposed process has 53 observable attributes; 41 measurements of the processing variables and 12 manipulated variables. The simulations proposed by Downs and Vogel [6] include 22 datasets, one being related to the normal operation of the process and 21 related to different faults that need to be detected and classified; the faults are described in Table 10.2. Attributes are measured and recorded in a synchronous manner, but with different periods: three, six, and fifteen minutes; one manipulated variable is not recorded, though. Measurements are subject to Gaussian noise. The datasets are composed of 960 observations for each 52 recorded attributes related to the lowest granularity (i.e. 3 minutes); the

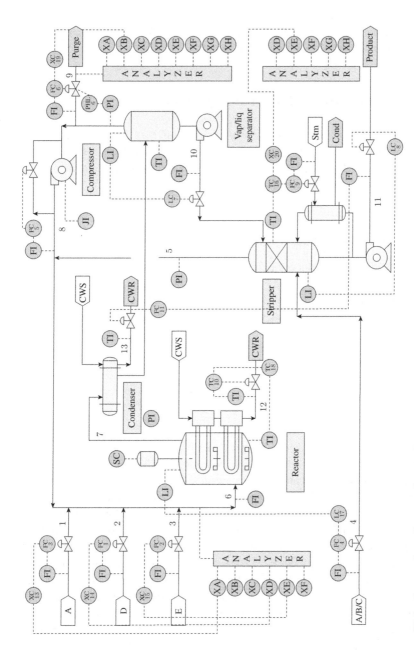

Figure 10.2 Process flow diagram of the TEP. Source: Adapted from [6].

Table 10.2 Faults in the TEP.

Fault	Process variable	Fault type
1	A/C feed ratio, B composition constant (Stream 4)	Step
2	B composition, A/C ratio constant (Stream 4)	Step
3	D feed temperature (Stream 2)	Step
4	Reactor cooling water inlet temperature	Step
5	Condenser cooling water inlet temperature	Step
6	A feed loss (Stream 1)	Step
7	C header pressure loss – reduce availability (Stream 4)	Step
8	A, B, C feed composition (Stream 4)	Random variation
9	D feed temperature (Stream 2)	Random variation
10	C feed temperature (Stream 4)	Random variation
11	Reactor cooling water inlet temperature	Random variation
12	Condenser cooling water inlet temperature	Random variation
13	Reaction kinetics	Slow drift
14	Reactor cooling water valve	Sticking
15	Condenser cooling water valve	Sticking
16	Unknown	Unknown
17	Unknown	Unknown
18	Unknown	Unknown
19	Unknown	Unknown
20	Unknown	Unknown
21	The valve for Stream 4 was fixed at the steady-state position	Constant position

Source: Based on [10].

attributes with a higher granularity consider the latest sampled value to fill the missing points in the time series.[1] Each time series has 960 points (which is equal to $960 \times 3 = 2880$ min $= 48$ hours so that the full data set has a size 52×960.

10.2.2 Tennessee Eastman Process as a Cyber-Physical System

Let us consider the TEP as a CPS following the procedure described in Chapter 7; remember that the TEP is a simulation (conceptual) model that represents a typical process of an industrial plant.

---

1 The TEP datasets are available at https://github.com/camaramm/tennessee-eastman-profBraatz (last access: September 7, 2021).

PS (a) Structural components: connections and valves; (b) operating components: reactor, condenser, vapor–liquid separator, recycling compressor- and stripper; (c) flow components: reactants A, C, D, E; (d) measuring components: sensors related to the different 52 attributes; (e) computing components: data acquisition module, data storage, and data processing unit (nodes connected by dashed lines); (f) communication components: cables connecting different components, and transmission and reception modules (dashed lines). The physical system is illustrated in Figure 10.2.

PF Generate products G and H through chemical reactions between A, C, D, E.

C1 It is physically possible to obtain G and H from chemical reactions described in Section 10.2.1.

C2 The process has to run without faults in the different stages. Twenty-one possible faults are described in Table 10.2, including the reactor or condenser cooling water inlet temperature and reaction kinetics, as well as unknown issues.

C3 For the conceptual model: a software that can simulate the TEP. For an actual realization of a TEP-like industrial process: availability of the reactants, personnel trained, the place where the physical system is to be deployed, maintenance investments, etc.

The three-layers of the TEP as a CPS are defined as follows.

- **Physical layer:** The schematic presented in Figure 10.2.
- **Measuring or sensing:** There are 52 attributes to be measured; the devices related to the measuring process are depicted by circles. The measures are synchronized with a granularity of three minutes, but some sensors have a larger granularity (6 or 15 minutes). The measuring is subject to Gaussian noise. Data are recorded as a numerical time series in discrete time.
- **Data layer:** It is not explicitly described where measured data are stored and then processed, fused, or aggregated. Consequently, the structure of awareness (SAw) cannot be defined.
- **Informing:** The processed data are then transmitted to a central unit that makes a decision (not presented in Figure 10.2).
- **Decision layer:** The decision is related to: (i) successfully detect the fault considering a given probability of a false alarm, and (ii) correctly classify the type of fault.
- **Acting:** No action is explicitly mentioned. Possibly, if a fault is detected, the process should be stopped by an agent Y. The structure of action (SAc) is then y.

The proposed framework presented in Table 10.1 can now be applied to help in defining our design problem: *How to build the data layer of the TEP?* We will follow

Table 10.3 Guiding questions for the TEP as a CPS as presented in [10].

Q#	Topic	Related question
Q0	Application	Detect faults.
Q1	Sensors	There are 52 measuring devices.
Q2	Samples	Periodic sampling with different granularities (3, 6, and 15 min); all variables are in sync (3 min).
Q3	Communication	Assumed perfect without explicit definition. **Open for design.**
Q4	Data storage	Assumed that the data of the sensors are recorded and available to be processed. **Open for design.**
Q5	Data fusion	(i) Principal component analysis (PCA), (ii) dynamic PCA, or (iii) canonical variate analysis (CVA).
Q6	Decisions	Q or T^2 statistical tests; details in [1].
Q7	Actions	Not defined; we may assume that the process stops whenever a fault is detected.

Source: Based on [10].

the seminal work [10] considering the fault detection problem and their respective solution. Specifically, questions 3 and 4 are not explicitly considered in the already mentioned TEP seminal works (Table 10.3). More recent contributions attempting to fill this gap study, for instance, different aspects related to industrial IoT communications, including the deployment of 5G networks [11] and the impact of missing data [12].

Fault detection requires monitoring of the process in real time, i.e. whenever new samples are acquired, they need to be transmitted and processed. Although three minutes (the lowest measuring granularity in the TEP) is a reasonably large value in the domain of communication in local data networks, and the decision procedure based on either the Q test or the T^2 test are not computationally complex, a suitable solution for fault detection in the TEP is a many-to-one topology considering wireless communication links from sensors to a gateway with computational capabilities for data processing and storage. Therefore, the proposed solution is a wireless radio access technology and edge computing keeping the fault detection process within the local (industrial) area network. Figure 10.3 illustrates this design, which is a simplified version of the proposal in [11].

For the fault classification, the data layer solution could be different. This type of problem is defined by complete data sets containing all time-indexed measurements. This is such an input given to the decision algorithm to classify if: (i) a fault has occurred and, if yes, (ii) which type. More advanced techniques for fault identification and then diagnosis could also be implemented in real time, also making

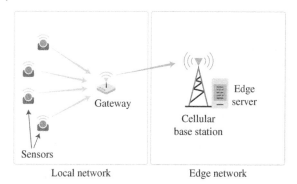

Figure 10.3 Proposed communication and storage architecture for fault detection in the TEP. Source: Adapted from [11].

explicit a step of expected interventions to be taken [1]. This would also lead to different requirements for the data transmission and processing, which is the focus of Exercise 10.1.

10.2.3 Example of Fault Detection in the TEP

After defining the TEP as a CPS, the attention will be devoted to exemplify the fault detection operation following the lines presented in Figure 10.1. As indicated in Chapter 9 when discussing machine learning methods, the fault detection requires a training phase where the statistical characterization of the normal process will be determined. With a full data set comprised of the time series of the 52 variables, the principal component analysis (PCA) method is used to fuse the data to reduce their dimensionality. With the reduced data set, a threshold for faults based on either the Q test or the T^2 test is defined considering a given level of significance that is related to an acceptable level of false alarms based on the normal operation data set. Once the threshold is defined, the fault detection can be done in real time using a rule-based approach that considers the following: a fault happens whenever the samples associated with the current measurements are above the threshold for six times in a row.

Figure 10.4 illustrates the proposed fault detection in the TEP following [10]. Figure 10.4a, b present the time series of an arbitrarily chosen variable in normal operation and when Fault 5 occurs (indicated by a circle), respectively. It is visually easy to see a considerable change in the time series behavior after the fault occurs at the time index 160. However, the proposed method for fault detection consists of a statistical test based on the fused data, which include the other 51 observable attributes. Figure 10.4c, d provide a visualization of the decision process for the Q and T^2 tests, respectively. The first interesting fact is that to avoid false alarms resulting from random variations, the proposed decision rule considers subsequent states (in this case, six), and thus, the fault detection is always subject to a minimum delay (or lagging time). The second aspect is the difference between

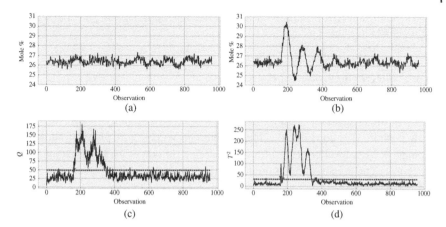

Figure 10.4 Fault detection in the TEP considering Fault 5 in Table 10.2. The first star indicates when the fault occurs. (a) Reactor feed analysis of B (stream 6): normal operation. (b) Reactor feed analysis of B (stream 6): Fault 5. (c). Q test. The second star indicates when the fault is detected. (d) T^2 test. The second star indicates when the fault is detected.

the two methods, the Q test being better than the T^2 to detect this specific fault because the former detects the fault in the observation 165 while the latter in the observation 173.

This case is only a simple example of the TEP, which is still a relevant research topic for testing the performance of fault detection methods. Currently, there are several research directions that focus on different aspects, namely data acquisition (e.g. event-triggered acquisition), advanced machine learning methods for detection, realistic models for data communications including missing samples, and data imputation for estimating missing samples to build a complete data set. Despite these advances, the most fundamental challenges of the TEP are still found in [1, 10]. The approach taken in this section provides a systematic framework to study the different opportunities of the TEP as a CPS, which includes both its aforementioned seminal results and the most recent ones as in [11, 12].

10.3 Cyber-Physical Energy System

Energy systems are large-scale infrastructures, whose operational management requires coordination of different elements. Specifically, modern electricity power grids are complex networks, whose operation requires a balance between supply and demand at a very low time granularity [7, 13]. This section deals with a specific intervention related to a proposal for distributed frequency control using smart fridges. Before starting this analysis, a very brief description of the power grid as a system will be provided next.

10.3.1 Electricity Power Grid as a System

The electricity power grid is an infrastructure whose main objective is to transfer energy in the electric form from one place where the electricity in generated to another where it is consumed. The grid refers to its complex physical network topology. A high-level concept of a modern electric power grid is presented in Figure 10.5. The demarcation of the power grid as a system is proposed next.

PS (a) Structural components: cables, towers; (b) operating components: transformers, power electronic components; (c) flow components: electric current; (e) computing components: data acquisition module, data storage, sensors, and data processing units; (f) communication components: cables connecting different components, and wireless devices.

PF Transfer electric energy from a generating point to another consuming point.

C1 It is physically possible to accomplish the PF.

C2 The power transfer needs to occur so that the supply and demand are balanced. Components have to be maintained and personnel trained, and an operation center has to detect faults and react to them, etc.

C3 Weather conditions, long-term investments, availability of supply, existence of demand, etc.

The three-layers of the power grid as a CPS are defined as follows.

- **Physical layer:** The physical components of the grid.
- **Measuring or sensing:** Several measuring devices for power grids to obtain the instantaneous values of, for example, current, voltage, and frequency.
- **Data layer:** It might be deployed at different levels: large data centers from the grid operator for near real-time operation, cloud computing for data analysis and trends, edge computing for quick reactions, and local processing without communication.
- **Informing:** The communication process is usually two-way from sensors to operators, and from operators to actuators. This depends on the level of operation that the solution is needed.
- **Decision layer:** This is related to decisions at different levels: it might be related to fault detection and diagnostic in the connected grid, or when to turn on a specific appliance.
- **Acting:** It can be from turning off a specific appliance in one household to shutting down a large part of the connected grid to avoid a blackout. This can be automated or not.

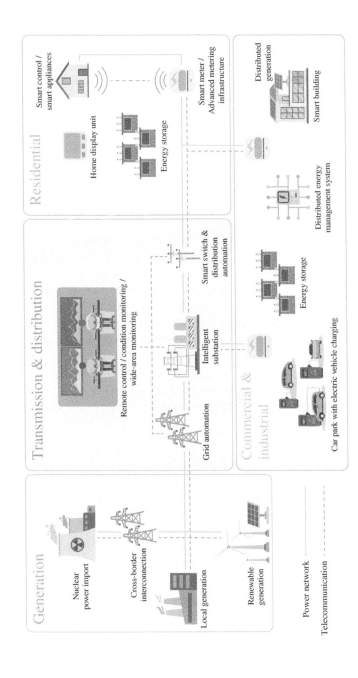

Figure 10.5 Illustration of a power grid.

10.3.2 Frequency Regulation by Smart Fridges

One of the main problems in the power grid is the frequency regulation. Alternating current power grids work synchronously with a given nominal frequency (usually 50 or 60 Hz). This means that all elements connected to the grid will experience the same frequency, which is related to the instantaneous balance between the energy supplied by the generators (mostly electric machines) that convert some form of energy into electric energy and the electric energy demanded by different loads. Then,

- if supply = demand, then the frequency is constant;
- if supply > demand, then the frequency increases;
- if supply < demand, then the frequency decreases.

Small perturbations are allowed within a given range. When these upper and lower boundaries are crossed, the grid operation requires an intervention in order to bring the system back to its desirable state. This is usually done by adding or removing generators on the supply side.

There is also the possibility to handle this issue on the demand side, considering two facts: (i) all elements in the grid experience the same synchronized frequency, and (ii) some appliances have demand cycles that can be advanced or postponed (considering the short time frame related to frequency regulation) without affecting their function. The proposed solution is to have fridges that react to the frequency signal by postponing or advancing they cooling cycles, thereby helping to restore the frequency to its desired operation. Table 10.4 states the questions to define the design problem of these smart fridges as part of a cyber-physical energy system.

Table 10.4 Smart fridge design in a cyber-physical energy system.

Q#	Topic	Related question
Q0	Application	Frequency control
Q1	Sensors	Observe the frequency
Q2	Samples	Periodic sampling
Q3	Communication	No communication
Q4	Data storage	Local storage for a few samples
Q5	Data fusion	No
Q6	Decisions	Is the frequency within the acceptable limits?
Q7	Actions	Reaction to frequency outside the limits. **Open for design**

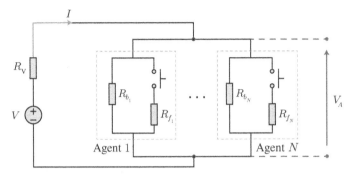

Figure 10.6 Direct current circuit representing the physical layer of the system. The circuit is composed of a power source with the voltage V and its associate resistor R_V, and several resistors in parallel, controlled by N agents. Each agent i has a base load R_{b_i} and a flexible load R_{f_i} that can be used in demand-side management. The number of active flexible loads determine the current I. Source: Adapted from [2].

To isolate the impact of different designs of actions that smart fridges could take, a simple model was proposed in [2] based on a simple direct current circuit model. Figure 10.6 presents the proposed simplified model. The rationale behind it is the following: A balanced supply (modeled as a voltage source V and an associated resistor R_V) and demand (modeled as N different agents representing households with a base load R_b and a flexible load R_f) leads to a voltage V_A within the operational limits. This model captures the effects of adding and removing loads in AC grids, where frequency is the operational attribute to be kept within the limits $V_{A,\,low}$ and $V_{A,\,up}$.

The dynamics of the circuit without smart fridges consists of individual fridges being on and off randomly, emulating cooling cycles. Measurements and actions are performed every second (lowest granularity). Four types of smart fridges are tested. The individual behavior of each type of smart fridge is described below.

- **Smart 1:** If $V_A[k] > V_{A,\,up}$, then the flexible load is off in time $k + 1$. If $V_A[k] < V_{A,\,low}$, then the flexible load is on in time $k + 1$. Otherwise, normal (random) behavior.
- **Smart 2:** If $V_A[k] > V_{A,\,up}$, then the flexible load is off in time $k + 1$ with a given probability p_{off}. If $V_A[k] < V_{A,\,low}$, then the flexible load is on in time $k + 1$ with a given probability p_{on}. Otherwise, normal (random) behavior. This is similar to the ALOHA random access protocol in wireless communications.
- **Smart 3:** If $V_A[k] > V_{A,\,up}$, then the flexible load will await a random time X to check to become off so that it is off in time $k + x$, where x is a specific realization of the random variable X. If $V_A[k] < V_{A,\,low}$, then the flexible load will await a random time Y to check to become on so that it is off in time $k + y$, where y

is a specific realization of the random variable Y. This is similar to the CSMA random access protocol in wireless communications.

- **Smart 4:** If $V_A[k] > V_{A, up}$, then the agent estimates from $V_A[k]$ the numbers of fridges $N_{off}[k + 1]$ that are required to be off, and thus, the flexible load is off in time $k + 1$ with a given probability $p_{off}[k + 1]$ that is a function of $N_{off}[k + 1]$. If $V_A[k] < V_{A, low}$, then, the agent estimates from $V_A[k]$ the numbers of fridges $N_{on}[k + 1]$ that are required to be off, and thus, the flexible load is off in time $k + 1$ with a given probability $p_{on}[k + 1]$ that is a function of $N_{on}[k + 1]$.

To test those different solutions, a simple numerical simulation is presented in Figure 10.7 with a setup arbitrarily chosen. The experiment consists of an intentional voltage drop between 1500 and 2000 seconds, which generates $V_A[k] < V_{A, low}$. The self-developing dynamics of the CPS is dependent on the type of reaction assumed by the smart fridge, considering the normal case as our benchmark.

What is interesting to see is that Smart 1 and Smart 2 designs are not adaptive, and they are hard wired in a way that instead of helping, the cycles of the smart fridges become synchronized, which results in further instabilities (worse than before, because now the system swings above and below the desired operation values). Smart 4 is an improvement of Smart 2 by adapting the internal activation or deactivation probability to the estimated state of the system, considering that all other fridges are acting in the same manner. Smart 3, in its turn, offers a way to desynchronize the reactions, avoiding the harm of the undesired collective behavior. It is easy to see this by comparing the voltage behavior and the number of active flexible loads.

In this case, the SAc of the CPS is equal to $\sum_{i=1}^{N} a_i$, because there are N agents $(A_1, \cdots, A_N)$ that can control their own flexible loads. What is more interesting in this study case is the SAw: (i) the benchmark scenario and Smart 1 have $\sum_{i=1}^{N} a_i$ and (ii) Smart 2, Smart 3, and Smart 4 have $\sum_{i=1}^{N} \left(\sum_{j \neq i} a_j \right) a_i$. The difference between the strategies in (ii) is how the images of the other elements are constructed. In Smart 2 and Smart 3, the images are static, and the randomization parameters in terms of the probability p and time X and Y are fixed. Smart 4, which visually presents the best outcome, considers an adaptive randomization where all appliances will collaboratively set their $p[k]$ based on an estimation of how many agents will activate their flexible load.

It is clear that this is a very specific and controlled study case, but it serves to illustrate the importance of predicting collective behavior in systems where the same resource is shared by different autonomous agents that both (i) cannot communicate to each other and (ii) observe the same signal that guides reactions. Despite its simplicity, the model captures well a phenomenon that might occur in a future where smart appliances react to universal signals (like open market price and frequency), but act without coordination. A detailed analysis of this is found in [2].

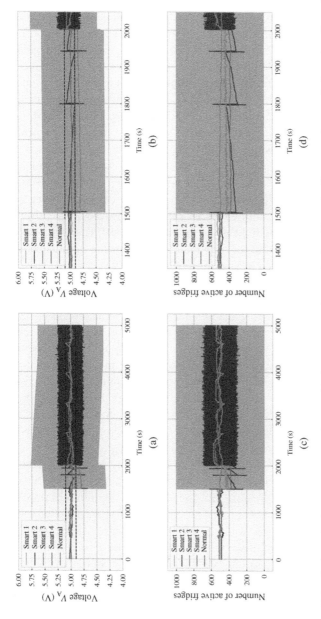

Figure 10.7 Self-developing system dynamics considering an external perturbation. There are five scenarios, one being a baseline where the fridges are not smart and four others representing scenarios where the fridges activate their flexible load based on different decision rules. (a) Voltage observed by the agents. (b) Voltage observed by the agents zoomed in. (c) Number of active agents. (d) Number of active agents zoomed in.

10.3.3 Challenges in Demand-Side Management in Cyber-Physical Energy Systems

This example indicates the potential benefits of demand-side management in cyber-physical energy systems. However, in real-case scenarios, the grid has a highly complex operation determined by intricate social, economical, and technical relations [14]. The main message is that, in actions related to demand-side management, ICTs have to be used to coordinate actions by creating a dynamic estimation of the grid state at the decision-making elements. Universal signals like the voltage in the toy example, or frequency and/or spot price in real world, may lead to undesirable collective effects that result in poor operational outcomes.

In the frequency or voltage control, the time of reaction ought to be low, and thus, the latency involved in the communication and computation might be prohibitive. In those cases, the local decision-making process with an associated action (i.e. the decision-maker and the agent are colocated at the same entity) as the one presented before is more suitable. This leads to a distributed decision-making process where the decision-making is local. Therefore, the CPS designer needs to consider that the autonomous elements are not independent, because they share the same physical infrastructure. In this case, the coordination mechanism should be internalized by the decision-making process of the individual elements in order to mitigate the risks of undesirable collective effects.

However, at larger timescales, demand-side management might also be related to actions associated with tertiary control with a time granularity of minutes [7]. It is at this level that the governance model assumed by the system operation is mostly felt. For example, market-based governance models are related to individual elements aiming at maximizing profits (or minimizing costs) in a selfish manner, which may lead to systemic instabilities that result in overall higher costs despite the individually optimality of the strategy taken [15]. There are possible solutions to this problem within the market arrangements, which include direct control of loads, different price schemes, and the use of other signals like colors [14–16]. Other option is to move away from overly complex competitive market governance models (always guided by competition with myopic selfish approaches of profit maximization or cost minimization) and organize the cyber-physical energy system as a commons to be shared, where the demand-side management would be built upon requests, preferences, and availability [9]; the benefits of this approach will be discussed in Chapter 11.

What is important here is to indicate that cooperation among the elements that demand electric energy may use communication links and more sophisticated computing methods to coordinate their actions because the operational timescales at tertiary control allow higher latencies, compatible with decentralized or centralized ICTs. Then, the technical challenges involved here are quite different from the ones in voltage or frequency control, although the fundamental phenomenon is very similar: uncoordinated or poorly coordinated reactions that result in collective behaviors that are harmful to the CPS self-development within its operational constraints.

10.4 Other Examples

In this section, two other cases will be presented in brief. They are: (i) a cyber-physical solution to provide a way to record and process data related to critical public health situations like pandemics and (ii) the undesirable effects of traffic route mobile applications. These examples will be further explored in Exercises 10.3 and 10.4.

10.4.1 Cyber-Physical Public Health Surveillance System

During the COVID-19 pandemics, several mitigation measures were taken to prevent the virus from spreading, mainly before the availability of vaccination. Several solutions rely on explicit interventions that consider the virus propagation process based on epidemiological models over networks (see Chapter 5), whose parameters were estimated based on data collected in several ways. This type of three-layer system is already familiar to the reader. Such a CPS is mainly designed to monitor the propagation of the virus, whose vector are humans. In this case, this would be a tool of surveillance to support public health.

As one may expect, this brings justifiable concerns in terms of privacy and reliability of the acquired data. One possible solution was proposed in [3], where the authors described a federated trustworthy system design to be implemented to support mitigation actions in future epidemics. The proposed architecture is presented in Figure 10.8, which is defined in [3] as:

> A federated ubiquitous system for epidemiological warning and response. An organic and bottom-up scaling at the global level is envisioned based on active citizens' participation. Decentralized privacy-preserving computations are performed from the edge to the cloud based on crowd-sourced obfuscated and anonymized data managed with distributed ledgers to empower trust. Incentive mechanisms for responsible data sharing align the public health mandate with citizens' privacy and autonomy.

The demarcation of this CPS is the task of Exercise 10.3.

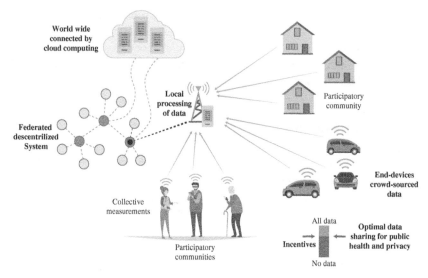

Figure 10.8 Cyber-physical public health surveillance system. Source: Adapted from [3].

10.4.2 Mobile Application for Real-Time Traffic Routes

The article [4] states an undesirable problem of mobile applications designed to help drivers to select routes, usually with the aim of minimizing the travel time. In the author's own words:

Today, traffic jams are popping up unexpectedly in previously quiet neighborhoods around the country and the world. Along Adams Street, in the Boston neighborhood of Dorchester, residents complain of speeding vehicles at rush hour, many with drivers who stare down at their phones to determine their next maneuver. London shortcuts, once a secret of black-cab drivers, are now overrun with app users. Israel was one of the first to feel the pain because Waze was founded there; it quickly caused such havoc that a resident of the Herzliya Bet neighborhood sued the company.

The problem is getting worse. City planners around the world have predicted traffic on the basis of residential density, anticipating that a certain amount of real-time changes will be necessary in particular circumstances. To handle those changes, they have installed tools like stoplights and metering lights, embedded loop sensors, variable message signs, radio transmissions, and dial-in messaging systems. For particularly tricky situations–an obstruction, event, or emergency–city managers sometimes dispatch a human being to direct traffic.

But now online navigation apps are in charge, and they're causing more problems than they solve. The apps are typically optimized to keep an individual driver's travel time as short as possible; they don't care whether the residential streets can absorb the traffic or whether motorists who show up in unexpected places may compromise safety.

The phenomenon described above follows a similar mechanism to Smart 1 of Section 10.3.2, where the individual decision-making process to the signal that is broadcast to all users leads to actions that result in an undesirable collective effect. In this particular case, however, there are a few differences because not all drivers use navigation apps, and not everyone who uses them employs the same type of app. Nevertheless, the way the algorithms are designed to determine routes in real time induces several drivers to select an commonly unused path, which from time to time leads to traffic jams and other related problems of overusing and/or underusing a shared infrastructure like streets and highways.

This engineered solution for decision support can be analyzed as a CPS, where the physical layer is the traffic infrastructure including cars, the data layer is a decentralized one, where data from street and highways are acquired and then processed by cloud computing, where informative data are produced to indicate the selected route, and the decision layer is constituted by drivers, who are the final decision-makers and agents who decide if the suggestion should be accepted and then act accordingly. It is interesting to compare how traffic jams, which are indeed unintended consequences of several cars selecting the same routes without explicit coordination, can also be formed by explicit interventions that are intended to reduce the travel time by avoiding busy routes.

Figure 10.9 illustrates this problem, described by the authors in the caption as [4]: *A sporting event at a nearby stadium [A] causes a traffic backup on the highway*

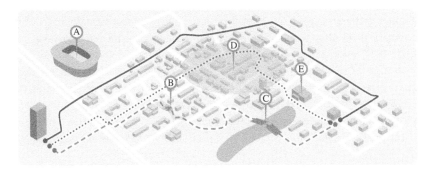

Figure 10.9 Illustrative example of undesirable consequences of navigation apps. Source: Adapted from [4].

that bypasses the center of this imaginary urban area. That's a problem for our hypothetical driver trying to get home from work, so she turns to a navigation app for help. The shortest–and, according to the app, the fastest–alternate route [dashed line] winds through a residential neighborhood with blind turns, a steep hill [B], and a drawbridge [C], which can create unexpected delays for those unfamiliar with its scheduled openings. The dotted line cuts through the city center [D] and in front of an elementary school [E]; the app doesn't know school just let out for the day. Fortunately, our driver knows the area, so she selects the full line, even though the app indicates that it isn't the fastest option. Drivers unfamiliar with the area and looking for a shortcut to the stadium could find themselves in chaotic–even hazardous–situations. A CPS designer that understands the fundamentals of collective behavior is capable of producing a solution to this problem; this is the task of Exercise 10.4.

10.5 Summary

This chapter presented different examples of CPSs to illustrate how the theory developed in this book is employed to study particular CPS cases. Two cases – an industrial CPS for fault detection and demand-side management in cyber-physical energy systems – were presented in more detail indicating positive and negative aspects that arise with the coupling between the physical and symbolic domains. Other two cases were left open for the reader as exercises. Nevertheless, the number of CPSs being currently deployed is constantly increasing, and thus, the reader can, of course, carry out a similar study considering the proposed eight questions and the demarcation based on the peculiar function with its respective necessary conditions of existence.

An important remark is that the study presented here did not focus on quantitative aspects but on qualitative ones. This decision is conscious and aims to highlight the impact of high-level design choices in contrast to narrower mathematical results, which are heavily dependent on the specific system setting. This latter aspect is obviously of key importance and is indeed the contribution of most references used here; the reader is then invited to read them to see how the quantitative tools presented in the previous chapters are applied in those papers (although they generally do not employ the CPS conceptual apparatus developed in this book). In the next chapter, we present the still missing discussion related to aspects of CPS beyond the technology itself.

Exercises

10.1 **Fault classification, identification, and diagnosis in the TEP.** In Section 10.2.2, a communication network solution was proposed for the fault detection problem where the disturbances need to be detected regardless of their types. This question is related to the (i) classification and (ii) identification & diagnosis problems.

(a) Explain the difference between a fault detection presented in Section 10.2.2 and a fault classification problem considering that the full data set is already available (i.e. the classification is not real-time).

(b) Based on (a), explain how the communication and storage design options change for the fault classification.

(c) Propose a design solution to the communication and storage considering the fault classification problem in the TEP.

(d) Propose a design solution to the communication and storage considering the real-time fault identification & diagnosis problem in the TEP following the box diagram presented in Figure 10.10.

(e) If the process recovery intervention depends on data transmission from the decision-maker element to possible different agents (actuators), explain the impact of this new data transmission on the data layer design.

10.2 **Industrial CPS and energy management.** The paper [17] proposes a CPS to acquire energy consumption data related to a specific industrial process. Figure 10.11 depicts the proposed solution to data acquisition of an aluminum process, including the representation of the physical, data and decision layers. The authors explain their approach as follows:

> Motivated by the automatic acquisition of refined energy consumption with [Industrial] IoT technology in [mixed manufacturing system], a supply-side energy modeling approach for three

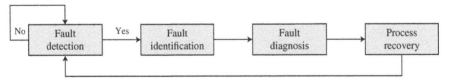

Figure 10.10 Process monitoring loop. Source: Adapted from [1].

types of processes is proposed. In this approach, a data collection scheme based on sensor devices and production systems in existing software is presented, see Figure 10.11 Modules (1) and (2). Three mathematical models are developed for different energy supply modes, see Figure 10.11 Module (3). Then, the production event is constructed to establish the relationship between energy and other manufacturing elements, including job element, machine element and process element. Moreover, three mathematical models are applied to derive energy information of the multiple element dimensions, see Figure 10.11 Module (4).

Without going into details of the process itself but referring to the original paper, the task of this exercise is to answer the guiding questions of this CPS presented in Table 10.5, in a similar way to the examples previously presented in this chapter.

10.3 **CPS for public health surveillance.** Read the paper [3] discussed in Section 10.4.1 and then demarcate the proposed cyber-physical public health surveillance system employing the same approach taken in Sections 10.2.2 and 10.3.1. Refer also to Figure 10.8.

10.4 **Suggestion of routes by mobile applications.** Propose a solution to the problem described in Section 10.4.2 using the ideas presented in Sections 10.3.1 and 10.3.3.

Table 10.5 Guiding questions related to Figure 10.11.

Q#	Topic	Related question
Q0	Application	
Q1	Sensors	
Q2	Samples	
Q3	Communication	
Q4	Data storage	
Q5	Data fusion	
Q6	Decisions	
Q7	Actions	

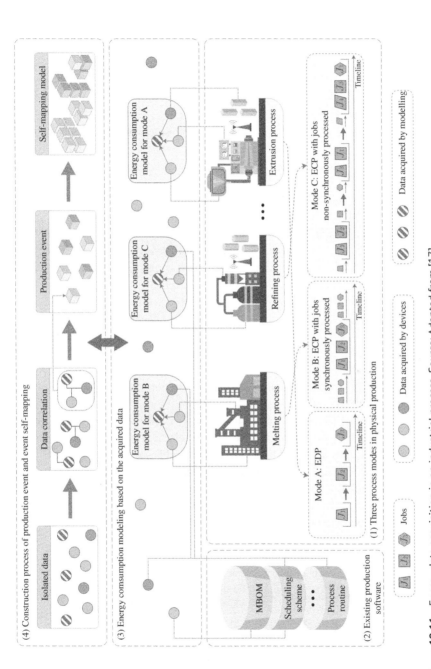

Figure 10.11 Energy data acquisition in an industrial process. Source: Adapted from [17].

References

1 Chiang LH, Russell EL, Braatz RD. Fault Detection and Diagnosis in Industrial Systems. Springer; 2001.

2 Nardelli PHJ, Kühnlenz F. Why smart appliances may result in a stupid grid: examining the layers of the sociotechnical systems. IEEE Systems, Man, and Cybernetics Magazine. 2018;4(4):21–27.

3 Carrillo D, et al. Containing future epidemics with trustworthy federated systems for ubiquitous warning and response. Frontiers in Communications and Networks. 2021;2. https://www.frontiersin.org/articles/10.3389/frcmn.2021 .621264/full

4 Macfarlane J. Your navigation app is making traffic unmanageable. IEEE Spectrum. 2019. https://spectrum.ieee.org/computing/hardware/your-navigation-app-is-making-trafficunmanageable

5 Gutierrez-Rojas D, et al. Three-layer approach to detect anomalies in industrial environments based on machine learning. In: 2020 IEEE Conference on Industrial Cyberphysical Systems (ICPS). vol. 1. IEEE; 2020. p. 250–256.

6 Downs JJ, Vogel EF. A plant-wide industrial process control problem. Computers and Chemical Engineering. 1993;17(3):245–255.

7 Nardelli PHJ, et al. Energy internet via packetized management: enabling technologies and deployment challenges. IEEE Access. 2019;7:16909–16924.

8 Hussain HM, Narayanan A, Nardelli PHJ, Yang Y. What is energy internet? Concepts, technologies, and future directions. IEEE Access. 2020;8:183127–183145.

9 Nardelli PHJ, Hussain HM, Narayanan A, Yang Y. Virtual microgrid management via software-defined energy network for electricity sharing: benefits and challenges. IEEE Systems, Man, and Cybernetics Magazine. 2021;7(3):10–19.

10 Russell EL, Chiang LH, Braatz RD. Fault detection in industrial processes using canonical variate analysis and dynamic principal component analysis. Chemometrics and Intelligent Laboratory Systems. 2000;51(1):81–93.

11 Hu P, Zhang J. 5G-enabled fault detection and diagnostics: how do we achieve efficiency? IEEE Internet of Things Journal. 2020;7(4):3267–3281.

12 Dzaferagic M, Marchetti N, Macaluso I. Fault detection and classification in Industrial IoT in case of missing sensor data. TechRxiv. 2021.

13 Nardelli PHJ, Rubido N, Wang C, Baptista MS, Pomalaza-Raez C, Cardieri P, et al. Models for the modern power grid. The European Physical Journal Special Topics. 2014;223(12):2423–2437.

14 Palensky P, Dietrich D. Demand side management: demand response, intelligent energy systems, and smart loads. IEEE Transactions on Industrial Informatics. 2011;7(3):381–388.

15 Kühnlenz F, Nardelli PHJ, Karhinen S, Svento R. Implementing flexible demand: real-time price vs. market integration. Energy. 2018;149:550–565.

16 Kühnlenz F, Nardelli PHJ, Alves H. Demand control management in microgrids: the impact of different policies and communication network topologies. IEEE Systems Journal. 2018;12(4):3577–3584.

17 Peng C, Peng T, Liu Y, Geissdoerfer M, Evans S, Tang R. Industrial Internet of Things enabled supply-side energy modelling for refined energy management in aluminium extrusions manufacturing. Journal of Cleaner Production. 2021;301:126882.

11

Beyond Technology

In Chapter 2, the definition of systems was introduced together with a method to specifically demarcate them in their peculiarity in order to guide the analysis of their particular (physical) realizations. At this point, the fundamentals of the theory of cyber-physical systems (CPSs) have been already presented, together with more practical examples and the description of the key enabling information and communication technologies (ICTs). What is still missing is the necessary articulation of the technical dimension as part of the social whole. The determination of a particular system by explicitly considering its conditions of reproduction and external conditions allows the CPS designer or analyst to define the level of effectivity of aspects beyond the technical development itself.

CPSs do exist materially in the world to perform specific functions. They affect, and are affected by, existing social practices; they may also create new practices, possibly modifying some aspects of society by reinforcing or challenging them. For example, during the decade that started in 2011, the growth of social media interactions, the widespread of high-speed mobile Internet and the development of algorithms for pattern recognition have produced changes in established political and economic practices (e.g. [1]). These effects are far from linear and can be seen as part of the different CPS self-developing reflexive–active dynamics that are relatively autonomous, but not at all independent of the world external to them. Although it would be impossible to provide a complete account, there are some aspects that are indeed more general and far-reaching, and thus, much more likely to produce real effects in the CPS design, deployment, and operation. This chapter critically discusses these aspects.

11.1 Introduction

It is unquestionable that we all live under the capitalist mode of production. Despite all ideological prejudice and political propaganda, Karl Marx is still

Cyber-physical Systems: Theory, Methodology, and Applications, First Edition. Pedro H. J. Nardelli.
© 2022 The Institute of Electrical and Electronics Engineers, Inc. Published 2022 by John Wiley & Sons, Inc.

fundamental to scientifically understand the complex exploitative dynamics of the modern society based on historically determined social forms. This theory was introduced in his magnum opus *Capital* [2], whose first words are:

> The wealth of societies in which a capitalistic mode of production prevails, appears as a 'gigantic collection of commodities' and the singular commodity appears as the elementary form of wealth. Our investigation begins accordingly with the analysis of the commodity.

Capitalism is a specific mode of production that is built upon the commodity form: the abstract concept that apprehends the reality of Capital; it is the elementary aspect of Marx's theory and of the actual capitalist social formations themselves [3]. Throughout thousands of pages, Marx produces a scientific theory (abstract theory) of society based on specific concepts and social forms (i.e. concrete abstractions), which serve as the scientific frame to assess any capitalist social formation. In particular, the text consists of several concrete examples of then existing social formations, notably England.

Against certain dogmatic and idealist readings of Marx, the philosopher Louis Althusser returns to the mature Marx of Capital considering it as the beginning of the *science of history*, i.e. the history punctuated by different dominant modes of production [4]. Although the discussions within the Marxist tradition and the critiques toward (academic and political) Marxism are very far from the scope of this book, I consider impossible to not move in this direction because the theoretical readings of Althusser, systematized mainly in [5], are the most suitable way to conceptualize the mutual effects between technology (including CPSs) and other aspects beyond the social whole.

In Althusser's words, the society is a *complex whole articulated in dominance* where the economic relation is the determination in the last instance. This means that, in the capitalist mode of production, the social phenomena are always affected by the commodity form, which organizes at some level the existing social relations and practices. In plain words, capitalism is a society of commodities where every single aspect of the world, not only including man-made things and processes but also the environment, animals, and humans with their labor power is a potential commodity to be produced, exchanged, exploited, accessed, or sold, aiming at the perpetual self-growth of capital (an unlimited abstract ending in itself). Note that this is not a reductionist argument, but rather acknowledges that in capitalism *money talks*.

This apparent digression is needed because CPSs are a result of the technical-scientific developments of capitalism, and thus, an articulation with the realities outside their boundaries must consider this fact. We actually have implicitly considered this when we have indicated, for example, that the existence of a given CPS requires maintenance, technical development, education, and trained personnel as conditions of reproduction or external conditions because, in capitalism, they are all related to the availability of funding, cost-benefit analysis, potential for profit, return of investment, and so on. Our role as a CPS designer or analyst is to properly articulate how those needs that are not technical in a strict sense are related to particular CPS technologies and their technical operation. One of the aspects is the governance model that implicitly or explicitly affects the design and operation of CPSs.

11.2 Governance Models

Governance is defined as [6]: *the act or process of governing or overseeing the control and direction of something (such as a country or an organization).* In 2009, the Nobel prize in economics was awarded to Elinor Ostrom *for her analysis of economic governance, especially the commons* [7]. Her main contribution is to prove that the usual economic classification that divides governance models into two mutual exclusive possibilities, namely the market-based model from the (neo)liberal theory and the state-based model based on the centralized planning optimization, is not sufficient and is indeed misleading [8, 9]. A commons-based governance model is then presented as an existing third option where specific norms are set by the users of a given shared resource themselves in order to define the way to manage its use.

Although most of her studies focused on special cases concerning communities and their respective environment to produce a social-ecological system [10], the categorization of governance models proposed by Ostrom is also suitable for the study of CPSs. In particular, CPSs are self-developing reflexive-active systems that operate under a specific governance model that might also determine the approach taken in their operation by affecting the internal decision-making processes. In the following subsections, a brief overview of the three (ideal) types of governance models and their impact in (and on) the operation of CPSs will be provided.

11.2.1 Markets

Market-based governance models presuppose that potentially everything (goods, services, persons, working power, etc.) can be *commodified*. In order words, they assume the existence of sellers who have property rights of the commodity X and buyers who are willing to spend money to acquire some rights of X. The price of X will be defined by the balance between the aggregate supply and demand of that commodity. The market relations are mostly studied by microeconomics supported by neoclassical economic theory, which strongly relies on mathematization.

Despite its elegance, such a theory is built upon very strong (unsound) assumptions of how the real interactions actually happen. These assumptions are that economic decision-makers/agents [11]: (i) can identify, quantify, and prioritize their individual choices without internal contradictions; (ii) always aim at maximizing their individual utility or profit; and (iii) are independent of one another and have access to all relevant informative data at the time of their (synchronous) decisions. The neoclassical theory mathematically proves that if those arbitrary axiomatic assumptions hold, the competitive market structure always leads to the most efficient allocation of X in the terms defined by the theory itself.

This result is the main strength and the main weakness of this way of theorizing social-economic phenomena. As a theory itself, its scientific value to apprehend the reality of economic processes is highly questionable because its main assumptions are empirically inconsistent with its object, despite the mathematical formalism and correctness. On the other hand, neoclassical economy provides a clear technical knowledge to guide and justify decision-making processes: it has a heuristic value to intervene in the reality it claims to produce the truth of. This interventionist rationality affects the design of policies and technologies alike.

A market-based governance model then shapes the conditions of reproduction and external conditions of CPSs through the assumption that CPSs are commodities themselves and/or produce commodities. In other words, the constitutive relations of CPSs are at some level mediated money (the universal equivalent) allowing exchanges of property rights determined by private law and contracts between free subjects of law; all these are valid and guaranteed by the state power. Note that property rights are different from, although related to, possession. The first depends on a third element – the state (or perhaps a blockchain) – to enforce the contracts between subjects of law (or of new ICT systems); property is then mediated and refers to the ownership guaranteed by a third party (or by a specific ICT system design). Possession, in turn, is related to the direct use or consumption of a specific thing, good, or process. The following example illustrates how the market-based governance model affects the operation of a specific CPS.

Example 11.1 *Cyber-physical energy system organized as markets* Consider the cyber-physical energy system described in the previous chapter. Its peculiar operation is to transfer electric energy from a given generating point (source) to a consumption point. We know that this is physically possible and the electric power grid is the system that performs this function. However, to actually deploy and operate this CPS, it is necessary to have specialized machines, equipment, and software, as well as engineers and technicians trained to work with this technical knowledge. If this CPS operates following a market-based governance model, there is a dependence on monetary investments that need to return, after some time, either as profits for private companies or competitive advantage for cities. These aspects are mainly classified as conditions of reproduction and external conditions, which affect the system operation but are usually associated with middle- to long-term time horizons.

With the introduction of ICT devices like smart meters and home energy management systems, some flexible loads might use the electricity price (which is defined by the electricity market mechanism) to take the operational decision of when to turn devices on or off. As discussed in Chapter 10, the existence of such a universal signal may lead to instabilities emerging from collective behavior [12]. As indicated in [13], a naive deployment of market signals to be used by smart devices to take operational decisions may (and probably will) create systemic instabilities. This may not only increase the electricity price as an unexpected effect but also harm the physical system operation. In this case, the assumption of independence and profit maximization that leads to efficient outcomes when subject to the same information meets the physical operation of the CPS. In summary, the market price as the main informative signal to be broadcast to the decision-makers (smart devices) leads to a poor operational outcome because it does not represent the actual operational state of the CPS; in other words, the market price *does not* (and cannot) reflect the actual balance between electricity supply and demand of the cyber-physical energy system on the desired operational time scale.

Although several problems have been reported (e.g. in California in the early 2000s and in Texas in 2021), market liberalization and commodification of electricity and associated services are seen as inevitable by a large majority of policymakers, researchers, and industries. Despite the relative success of the development of electricity markets in Europe, issues such as high prices, energy poverty, dependence on imports, and environmental pollution exemplify a few problems that are consequences of such (neo)liberal energy policies. Given the supremacy of neoclassical theory in university education, the usual solution given by the experts, think tanks, and lobbyists alike (which is considered the only possible solution) is to *unlock the value* and *create new markets*, which are widespread capitalist mottoes. Even progressive solutions like energy communities and electricity sharing are subordinated to market imperatives [14].

In the case of cyber-physical energy systems, the assignment of responsibilities for interventions related to different time scales (e.g. [15]) enables corrective actions like up- and down-regulation. This, on the other hand, may increase the electricity price, but that could be corrected by a more suitable type of market integration that considers what the electricity market price reflects as a signal [13].

Today, the market-based governance model is almost a natural choice because prices are always given and universal. In practice, the problems of the unsoundness of neoclassical theory are solved by particular institutional designs and technical interventions, as well as education to produce the rational human beings axiomatically defined by the theory. There have been several studies (including several Nobel prizes) that have tried to correct the assumptions of neoclassical theory; the reader is referred to the last chapter of [11] for some examples. One remarkable and very pragmatic approach is to consider the market economy as a multiagent complex system. The engineer–economist W. Brian Arthur is probably the best-known author in the field called Complexity Economics, whose main principles have been summarized in [16].

The heuristic value of Complexity Economics, heavily supported by the advances in computer sciences and ICTs, does not change the core of neoclassical theory. Markets are theoretically studied based on universal rules of behavior and concretely deployed through enforcement of contractual rules (or private law). The freedom of market relations is bounded by those types of rules that define the possibilities that individuals have (even if they are heterogeneous and have access to different sources of information); in this sense, individuals become subjects to the given rules (or to the law). Because individuals cannot (directly) deliberate about (abstract) rules that affect their relations with each other, their autonomy is (strongly) bounded. Nevertheless, there is a certain level of autonomy where individual decision-makers are usually agents, making the market-based governance model associated with a distributed topology of decision-making process and action. In the following section, we will move to another governance model related to the centralized approach of planning, where the autonomy in decision-making is further limited but, counterintuitively, opening the possibility for more effective individual and social outcomes.

11.2.2 Central Planning

In contrast to the market-based governance model where decision-makers are considered independent and formally equal in relation to the rules, central planning is fundamentally grounded in an asymmetry in the decision-making process. In this case, there is one central decision-maker that sends direct commands to agents. This centralized approach provides a straightforward way to coordinate agents because the decision-maker can produce a schedule of actions to be taken by them. Central planning usually refers to disciplines of operational research and systems engineering [17].

Such a governance model is, nonetheless, fully integrated into the capitalist mode of production and the commodity production. In fact, the Fordist regulation of the period usually called golden years of industrial capitalism has been strongly related to centralized planning and hierarchical decision-making [18].

Despite its key differences from the post-Fordist regulation (which is related to decentralization of operations toward a market-based governance model), central planning is not opposed to the capitalist social forms but it assumes a rigid technocratic distinction between the elements of the system: one capable of "thinking" and planning (the decision-maker) and several others of "doing." A one-to-many topology where one element controls and all others follow its direct orders directly emerges from that.

In CPSs, this is reflected in situations where there is only one element that is a decision-maker orchestrating actions in order to reach the optimal operation of the system considering some performance metric to be optimized. This approach is usually operationally driven, and thus, the effectivity of the CPS to perform its peculiar function is put at the center of the optimization problem. This generally leads to maximization problems related to a utility that captures the efficiency of the CPS to accomplish its operational function. It is also possible that, like in market-based governance models, price signals are used to make decisions and guide actions, aiming at profit maximization or loss minimization. In this situation, however, the market price is given as an external signal, and the goal of price is not to balance supply and demand that are internal to the CPS (in contrast to the market-based governance approach when it is internally used by distributed decision-makers). In the following, an example of central planning in a cyber-physical energy system will be provided.

Example 11.2 *Central planning in cyber-physical energy systems* In Example 11.1, individual smart devices reacted to price signals without coordination, which resulted in an unintended collective effect that produced instabilities in the grid operation. If a governance model based on central planning is considered, a decision-maker element (e.g. load controller located at the distribution grid operator) will directly control which devices (agents) will be turned on and off in a specific time [15]. This decision is usually taken after solving an optimization problem whose objective function is related to operational aspects of the grid, such as cable capacity, availability of supply, and reactions to faults.

Such an approach may also aim at profit maximization if the grid operator is a private company. In this case, the operational aspects are defined by law. The objective function then includes compensation fees and other sanctions that are imposed by energy regulatory agencies when the system operates outside such contractual or regulatory limits.

A centralized approach is also taken by large power plants that prepare to produce a certain amount of electric energy in specific periods. The schedule of the operation of the plant is defined by one decision-maker to dispatch to the grid the electric energy needed or legally agreed upon.

Despite the differences between the two governance models, this example indicates that, in the capitalist mode of production, central planning is subordinated to market imperatives (e.g. maximizing profit or improving cost-efficiency). However, in central planning, price signals are used only as an external input for the optimization procedure, which organizes the agents' interventions. In any case, depending on the CPS under consideration, a central planning governance model might be used to prioritize operational aspects of the system self-development rather than internalize price signals as information to be used by distributed decision-makers to decide about their actions, creating the possibility of collective effects. On the other hand, network topologies that rely on central elements are more vulnerable to targeted attacks.

Nevertheless, most large-scale CPSs like cyber-physical energy systems and industrial CPSs are neither completely distributed nor completely centralized in their decision-making processes. Therefore, they can be assessed by systematically studying their own peculiar functions and the complex articulations between the specific CPS and the environment in which it exists, which is also composed of other systems (cyber-physical or not). As the two examples presented so far indicate, in capitalism, the articulation between the CPS and the economy shall always be considered, either operationally or as an external condition. Following the groundbreaking empirical research of Elinor Ostrom [8] about the commons and the theoretical contributions by Karl Marx [2], the next subsection focuses on a different mode of production that exists at the margins of capitalism but that offers an alternative to the commodity form following commons-based peer production [19].

11.2.3 Commons

The commodity form as a universal abstraction necessarily supported by the state power and by a legal system nucleated around private law is very recent in history [2, 19]. In 2021, the tendency of commodification of every possible aspect of reality is probably stronger than ever, enclosing not only physical spaces but new cyber-physical ones. Opposing it and, at the same time, opposing a centralized state power to command the access of shared good, processes, and environments (being them physical or cyber-physical) [9], Ostrom et al. studied the commons as a type of governance used in several communities that shared common resources (e.g. fisheries); her work is so remarkable that she won the Nobel prize in economics for her studies.

The term commons refers to a shared resource that needs to be managed to avoid overutilization, which may harm its existence. In her empirical study of local communities, Ostrom showed peculiar institutional arrangements produced by the members themselves to (self-)govern the shared pool of resources, summarized in eight principles [Table 3.1][8] (with modifications indicated by [·]):

(1) Clearly defined boundaries (…)
(2) Congruence between appropriation and provision rules and local conditions (…)
(3) Collective-choice arrangements [where] most individuals affected by the operational rules can participate in modifying the operational rules.
(4) Monitoring (…)
(5) Graduated sanctions [where] appropriators who violate operational rules are likely to be assessed graduated sanctions (depending on the seriousness and context of the offense) by other appropriators, by officials accountable to these appropriators, or by both.
(6) Conflict-resolution mechanisms [where] appropriators and their officials have rapid access to low-cost local arenas to resolve conflicts among appropriators or between appropriators and officials.
(7) Minimal recognition of rights to organize [where] the rights of appropriators to devise their own institutions are not challenged by external governmental authorities.
(8) Nested enterprises [for larger systems where] appropriation, provision, monitoring, enforcement, conflict resolution, and governance activities are organized in multiple layers of nested enterprises.

Ostrom's approach is empirical in the sense that it identifies institutional principles that exist in communities that share a resource among their members. Her work is also normative because these principles are associated with appropriation rules that govern a common pool of resources (i.e. the commons) without being guided by market-based or central-planning-based governance models. More interestingly, such an institutional-normative solution is not universal and depends on the resource under consideration; it is also based on possession and not on private property or a universal legal system. In other words, the commons is, in principle, not subordinated to the commodity form. Despite its importance, Ostrom did never foresee it as the basis of a new mode of production.

Other authors, in their turn, were motivated by her work to find alternatives to capitalism. Specially with the development of the Internet together with new forms of peer production and distribution of cyber goods such as Wikipedia, Linux, and BitTorrent, the potential of new more participatory technologies was identified as the seed of a new society [20]. From those ideas, Bauwens et al. defended in [19] a new mode of production based on peer production and peer-to-peer (P2P) exchanges where resources are shared in a common pool. In their own words:

P2P enables an emerging mode of production, named commons-based peer production, characterized by new relations of production. In commons-based peer production, contributors create shared value through open contributory systems, govern the work through participatory practices, and create shared resources that can, in turn, be used in new iterations. This cycle of open input, the participatory process, and commons-oriented output is a cycle of accumulation of the commons, which parallels the accumulation of capital.

At this stage, commons-based peer production is a prefigurative prototype of what could become an entirely new mode of production and a new form of society. It is currently a prototype since it cannot as yet fully reproduce itself outside of mutual dependence with capitalism. This emerging modality of peer production is not only productive and innovative 'within capitalism,' but also in its capacity to solve some of the structural problems that have been generated by the capitalist mode of production. In other words, it represents a potential transcendence of capitalism. That said, as long as peer producers or commoners cannot engage in their self-reproduction outside of capital accumulation, commons-based peer production remains a proto-mode of production, not a full one.

As indicated above and exemplified by the very examples of the Internet (Wikipedia, Linux, and BitTorrent) or more recent ones like Bitcoin, commons-based peer production, and P2P exchanges can be (and are) subordinated by the commodity form as far as the dominant mode of production is (still) capitalism. However, the proposed concept of *cosmolocalism* that would arise from the commons as a foundational social form based on peer production and shared appropriation of resources provides another way to design CPSs. In this case, price signals are excluded from the CPS self-developing operation while trying to isolate it from the commodity form. In other words, whenever possible, the CPS design and operation shall avoid market relations that should be kept at the minimum level as part of its external conditions. An example of a commons-based cyber-physical energy system as proposed in [21–24] will be presented next.

Example 11.3 *Energy Internet as a CPS governed as a commons* Inspired by the way Internet operates (see Chapter 9) using a packet switching technology, some researchers proposed a new approach to electricity grids denominated *packetized energy management* (PEM), which may serve as the basis of the Energy Internet [21, 25]. Its basic idea is to handle noncritical, flexible loads considering energy packets of x Wh during a period of T minutes (i.e. a discretized version of the electric energy demand), as illustrated in Figure 11.1. The proposed solution based on PEM is that each flexible device, such as a washing machine, space heating, and electric vehicle (EV) charging would send use requests explicitly

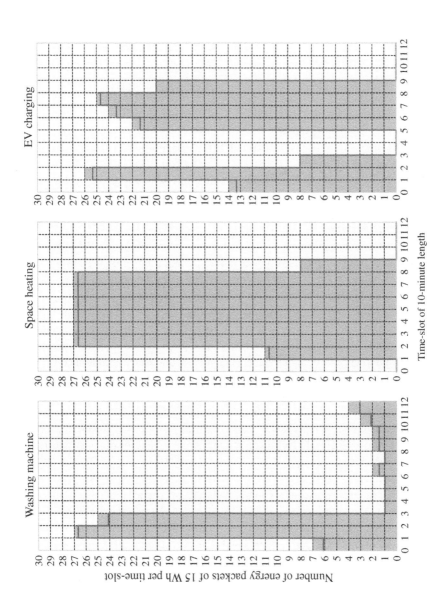

Figure 11.1 Illustration of three different packetized electricity demand profiles. Source: Adapted from [23].

indicating the desired quality of service (e.g. the EV should be charged at 6 am, or the living room is to be at 24 °C at 5 pm); following the nomenclature of the Internet, these flexible loads are *energy clients*. The requests are aggregated and prioritized based on certain rules to perform demand-side management, where the *energy server* determines which requests can be served, and for those that are selected, a schedule for their activation. The client server architecture of PEM resembles the Internet, and thus, the cyber-physical energy system that handles flexible loads in such a way should be called the Energy Internet.

The solution of the resource allocation problem is combinatorial and related to a known optimization problem called *flexible job shop scheduling* [26]. The difference here is that the client-server approach is a multiagent one and based on two steps: a bottom-up (the requests) and a top-down (the schedule proposed by the server). In addition, the procedure should be dynamic. In this case, the most suitable approach is using what it is called collective learning [27] that could be implemented through the *Iterative Economic Planning and Optimized Selections* (I-EPOS) system [28].

PEM and the Energy Internet can be seen as a hybrid of the market-based and the central-planning-based governance models, maintaining their key advantages. However, as indicated in [22], the Energy Internet could emerge in the future as a commons where electricity is peer produced by distributed energy resources like solar panels, considering also the possibility of short- and mid-term storage by heat pumps and batteries. The rules to be used by the energy server shall be jointly defined by the group of users, and all energy produced shall be shared in a common pool without being produced, consumed, and exchanged as a commodity. In this case, the Energy Internet would be governed as a commons following the already mentioned principles proposed in [8]. This means that energy produced by all Energy Internet members is (virtually) aggregated to be shared (not traded) based on needs regulated by jointly defined norms, which are always open for deliberation. The main technical and social challenges of this solution have been described in [23, 24].

The most interesting aspect of commons-based governance models is the decentralized, yet structured, nature of the decision-making process focusing on the use or appropriation – and not the property – of a given shared or peer-produced resource by agents considering their needs. On the other hand, large-scale CPSs governed as a commons need all types of support, such as technical, financial, and scientific, to be able to sustain their existence at the margins of capitalism. Nevertheless, as pointed out in [19], such designs might be part of a new society based on sharing, constituting then a new mode of production based on the commons. In this case, the capitalist mottoes would change to *unlock the access* and *create a new commons*.

11.2.4   Final Remarks About Governance Models

This section provided a very brief review of three types of governance models and how they can potentially affect CPSs. It is noteworthy that not all CPSs are equally impacted by the governance model, but they are all at some level subjected to it. We have illustrated three examples based on cyber-physical energy systems where market-based, central-planning-based, and commons-based approaches are internalized in the system self-development. The objective is to indicate the importance of the governance models when analyzing and designing CPSs, not only neglecting them or considering them as something irrelevant or independent. The core argument is that the governance model is both (implicitly or explicitly) constitutive of the CPS design and (indirectly or directly) articulated with its operation. The reader is invited to think more about it in Exercise 11.1.

11.3   Social Implications of the Cyber Reality

This section presents different emerging domains where the cyber reality, which is intrinsic of CPSs, has been disruptive in social terms. The topics covered are: data ownership, global platforms, fake news, and hybrid warfare. They are clearly interwoven, but it is worth discussing them individually and in this order. The rationale is that data regardless of the legal definition of their *ownership* are usually associated with *global platforms* that possess hardware capabilities for data storage and processing in huge data centers, cloud servers, and super computers. Processed data usually produce semantic information, which can be manipulated to become misinformation – part of this phenomenon is today (in 2021) called *fake news*. The widespread of fake news is analyzed by several authors as part of military actions, creating *hybrid warfare* associated with geopolitical movements.

11.3.1   Data Ownership

As mentioned in the previous section, capitalism depends on private property defined by legal rights, in contrast to direct appropriation and possession. When this is related to a concrete thing X, it is easier to indicate the ownership. For instance, a house X, which is in possession of a person Y, is legally owned by a bank Z; Y is allowed to use X following the contract signed with Z. The cyber domain brings new challenges: who is the owner of data (as a raw material) and information (as processed data)? Are data owned by the one who acquires them? Or the one who legally owns the hardware used in the data acquisition? Or even the one who stores and processes the data? The following example illustrates these challenges.

Example 11.4 *Data ownership of a video* Consider this situation. A given person is used to stay in the terrace to watch the movements of their neighbors, memorizing the time that every person leaves and arrives home. The person uses a paper notebook to record the times. Despite being weird, the notebook is, in principle, in the ownership of that person, and so are the data recorded in there. Now, the same person decides to install a digital camera at the same place he is used to observe the neighbors. The camera enables a video stream and the person could continue watching/following the neighbors but now comfortably on the sofa. Although the person is still writing in the notebook, the video is recorded by the company that sold the camera service, which is a surveillance company that shares anonymous data with another department of the company that offers private security to neighborhoods. The data of the video are stored in cloud data centers of large global companies, which use the metadata (where and when the video is recorded) to offer better business contracts to the companies. The data are used by several elements (legal entities) in this chain, but it is unclear how to define the ownership in this situation.

This example points to the challenges that the cyber reality brings in a mode of production where informative data are commodified, and in this case, can be traded after being processed. Thus, a definition of data ownership becomes necessary. This is, in fact, an ongoing discussion, and different national states are legislating their own solutions. So far in 2021, the most remarkable legal definition is probably the *General Data Protection Regulation* (GDPR) [29, 30] defined by the members of the European Union. Without discussing the details, GDPR aims at offering more rights to legally defined individuals to control their data in order to facilitate business operations of corporations in a world increasingly composed of CPSs. Although this, in principle, complies with the idea of empowering people's control over their data, the actual results might be different because most data processing power is located in very few profit-driven global platforms.

11.3.2 Global Platforms

There is only a handful of private enterprises that own large-scale platforms (i.e. hardware – data centers and super computers – logically connected through the Internet) that are employed to store and process the (big) data acquired worldwide [31]. Furthermore, these platforms offer widely used services and applications like emails, calendars, and text editors and, at the same time, enable third parties (being them other companies, governmental agencies, or individuals) to run their own applications. As their name already indicates, platforms are deployed to serve as a shared infrastructure (forming, in the current business language, ecosystems) where different applications coexist and run. Therefore, they are nowadays essential for CPSs to exist and operate.

The importance of such global platforms in the world of 2021 is unquestionable, and their economic, social, and political powers are growing steadily (the reader is referred to [32], for example). There are already uncountable issues related to political interventions, prejudice, data leak, and service outages. The strong dependence of fundamental aspects of society – including applications related to CPSs – on very few for-profit platforms is problematic. The book [33] presents several examples of such problems; actually, the main claim is that they tend to be a rule rather than only exceptions. Some other issues have been reported in [34], considering personal assessments during the growth of one of the largest global platforms that exist at the moment when this book is being written. One remarkable problem of today's Internet very related to social media controlled by such platforms refers to the phenomenon of fake news.

11.3.3 Fake News

Traditional media channels such as newspapers, radio, and TV have been (correctly or incorrectly) considered reliable sources or curators of news (or in our wording, informative semantic data). With the widespread of the Internet and the emergence of platforms, every user is a potential producer of content, including data that might be (intentionally or not) incorrect, i.e. a source of misinformation or disinformation. This characterizes the context in which the term "fake news" has been defined. The reference [35] presents a definition of fake news that is reproduced next.

> We define "fake news" to be fabricated information that mimics news media content in form but not in organizational process or intent. Fake-news outlets, in turn, lack the news media's editorial norms and processes for ensuring the accuracy and credibility of information. Fake news overlaps with other information disorders, such as misinformation (false or misleading information) and disinformation (false information that is purposely spread to deceive people).

Fake news concern the domain of semantic data, bringing more uncertainty in existing structures of meaning. The key problem is that, as mostly discussed in Chapters 3 and 4, uncertainty can be only studied with respect to some givens (i.e. well-determined structures of assumptions, which can vary, but these variations shall also be well determined). If this is not the case, then the uncertainty assessment becomes unfeasible to decision-makers. This phenomenon was described in detail in a recent report by Seger et al. [36], which also proposes ways of *promoting epistemic security in a technologically-advanced world*.

In this sense, the epistemic insecurity caused by disinformation attacks against stable structures of semantic information seems to be the function of fake news.

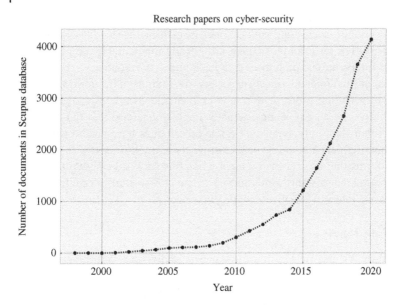

Figure 11.2 Number of publications in the Scopus database containing the term "cyber-security" or "cybersecurity".

However, it is very unlikely that this is something that has emerged spontaneously in society, but rather it seems to be closely linked to two interrelated trends of today's society, namely cyber-security (and the large business behind it) and hybrid warfare (and the geopolitical rearrangements after the Cold War, the 2008 economic crisis, and the emergence of global platforms among other relevant issues).

11.3.4 Hybrid Warfare

With the widespread of CPSs, security issues are increasingly moving from physical attacks toward cyber attacks. Just as an illustration, we have plotted the number of research articles published from 1980 to 2020 containing the words: "cyber-security" or "cybersecutiry." Figure 11.2 indicates an exponential-like growth, remarkably after 2015.

Despite the recent general interest, the research in cybernetics has always been associated with security and military applications since the Cold War years [37–39], leading to two traditions – one from the USA another from Russia – whose principles are remarkably similar.

In the USA, the cybernetics ideas for the military can be framed by the observe–orient–decide–act (OODA) loop described by United States Air Force Colonel John Boyd [40], which was influenced by the US-based cyberneticians

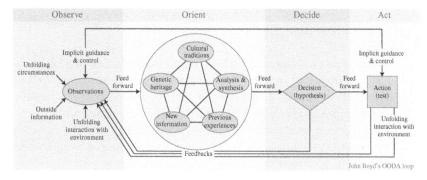

Figure 11.3 Boyd's OODA loop. Source: Adapted from: https://en.wikipedia.org/wiki/OODA_loop.

like Nobert Wiener. Figure 11.3 depicts the OODA loop, where observations provide informative data for orientation that are then used for decisions about actions. Note that the OODA loop approach is similar to the 3-layer CPS concept defined in Chapter 7. For Boyd, the warfare should involve operations in the OODA loop by, for example, injecting intentionally false semantic data (i.e. disinformation of a cyber attack) or sending electromagnetic signals to disrupt transmissions (i.e. physical cyber attacks). By purposefully affecting the OODA loop, the attacker might be capable of producing the desired results without physical battles, bombing, or other military interventions. The best analysis of the contribution of Boyd to contemporary military thinking is provided in [40].

On the Russian side, the research field is called reflexive control, defined as [41]: *a means of conveying to a partner or an opponent specially prepared information to incline him to voluntarily make the predetermined decision desired by the initiator of the action.* The idea of reflexive control started in the 1960s with [42], already discussed in Chapter 7, and it is still an active field nowadays developed as part of the *third order cybernetics* research activities [43]. The first military researcher outside the former Soviet countries to pay attention to this methodological way to approach warfare was the already retired Lieutenant Colonel Timothy Thomas from the US Army [41] in an attempt to better understand the military thinking in Russia after the Cold War. In a recent monograph, Major Antti Vasara from the Finnish Defense Force [44] revisited the Russian reflexive control in an attempt to bring more light to the different military operations (fairly or unfairly) imputed to today's Russia.

Although it is unfeasible to delve deeper into this topic, which is also blurry because of the nature of military conflicts themselves, the ICT-based, CPS-based warfare is pervasive, including relations to uprisings and conflicts worldwide. The term *hybrid warfare* is usually employed to characterize the nature of these conflicts. In a policy brief *What is hybrid warfare?* we read [45]:

> Various characteristics have been attributed to HW [hybrid warfare] conducted by non-state actors. First, these actors exhibit increased levels of military sophistication as they move up the capabilities ladder, successfully deploying modern weapons systems (like anti-ship missiles, UAVs), technologies (cyber, secure communication, sophisticated command and control), and tactics (combined arms) traditionally understood as being beyond the reach of nonstate adversaries. Combining these newly acquired conventional techniques and capabilities with an unconventional skill set – and doing so simultaneously and within the same battlespace – is seen as a potentially new and defining characteristic of non-state HW. This emphasis on greater military sophistication and capabilities is one of the key features of non-state actors using HW.
>
> A second core characteristic of non-state HW is the expansion of the battlefield beyond the purely military realm, and the growing importance of non-military tools. From the perspective of the nonstate actor, this can be viewed as form of horizontal escalation that provides asymmetric advantages to non-state actors in a conflict with militarily superior (state) actors. One widespread early definition of HW refers to this horizontal expansion exclusively in terms of the coordinated use of terrorism and organized crime. Others have pointed to legal warfare (e.g. exploiting law to make military gains unachievable on the battlefield) and elements of information warfare (e.g. controlling the battle of the narrative and online propaganda, recruitment and ideological mobilization).
>
> (...)
>
> The single critical expansion and alteration of the HW concept when applied to states is the strategically innovative use of ambiguity. Ambiguity has been usefully defined as "hostile actions that are difficult for a state to identify, attribute or publicly define as coercive uses of force." Ambiguity is used to complicate or undermine the decision-making processes of the opponent. It is tailored to make a military response—or even a political response–difficult. In military terms, it is designed to fall below the threshold of war and to delegitimize (or even render politically irrational) the ability to respond by military force.

It is clear that the description of hybrid warfare fits well with the development of both CPSs for military functions and for the use (and abuse) of the new opportunities that the cyber domain opens to new types of attacks. Disturbances in elections, military-juridical coups, and civil wars employing epistemic insecurity as a military tool are related to the new economic powers and geopolitical movements in the post-Fordist time we currently live in [1]. All these rearrangements in the economic and (geo)political power associated with the emergence of global platforms can be seen as a social impact of CPSs as part of the capitalist mode of production. The new technologies do work, but their effects beyond technology can lead to consequences that increase the exploitation of work and nature, reinforce existing oppressive relations, and support new types of military operations.

11.4 The Cybersyn Project

To end this chapter, a remarkable – and not well known – historical fact will be briefly introduced: the never realized cybernetic management of the Chilean economy. In the early 1970s, the democratically elected president in Chile, Salvador Allende, started the implementation of a distributed decision system to support the management of the Chilean national economy. The leader of the project was the recognized British researcher Stafford Beer, who first developed the field of cybernetic management by constructing the viable system model (VSM) [46]. Figure 11.4 illustrates a case of the VSM for the management of a socioeconomic scenario. The main idea is to incorporate both physical and information flows to operate and manage (sub)systems and their relations to viably perform their functions in a resilient way because the VSM is designed to be adaptive with respect to changes in the environment. In other words, the VSM sees (sub)systems forming as an organic, cybernetic whole (a system of systems).

Despite the active participation of unsuspected researchers, universities, and companies from the USA and England, the Cybersyn project had the same fate as president Allende: a victim of the Cold War and the fight against socialism in Latin America. The plan of creating a communication network and a decision support system to manage the Chilean economy in a distributed manner failed for reasons beyond technology. Chile paid the price of the geopolitics of the 1970s, and the ICT development to produce the social impact needed in Chile never took hold. Instead, in contrast to the economic plan for social development to produce a fair society, the military coup implemented (neo)liberal economic policies by force, making the Chile under the military dictatorship a laboratory of today's mainstream approach to the economic policies.

An important historical account in the complex relation between technology and other social processes is presented in [47]. The lesson to be learned is that technological development is far from being neutral, as already indicated by Marx [2] and further developed by Feenberg [20]. The design, operation, and actual

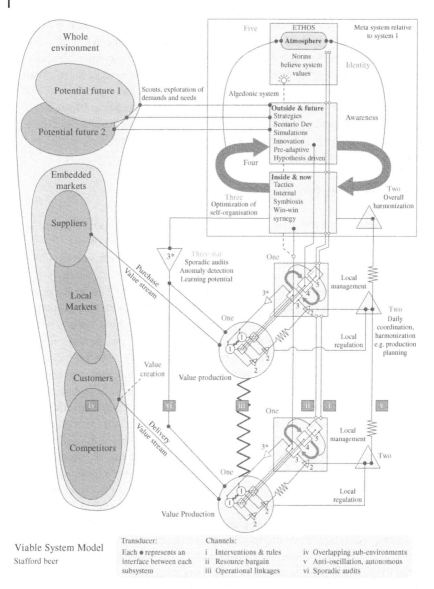

Figure 11.4 Example of Beer's VSM. Source: Adapted from: https://upload.wikimedia
.org/wikipedia/commons/b/b6/VSM_Default_Version_English_with_two_operational_
systems.png.

deployment of technical systems are always subject to power relations, mainly economic, but also (geo)political and social. The story of the Cybersyn project makes very clear the challenges ahead if one acknowledges that the vision of a society dominated by the (peer-produced) commons to be shared and managed by CPSs is, in many ways, similar to the cybernetic economic planning envisioned by Beer and Allende.

11.5 Summary

The theory constructed throughout this book allows the incorporation of aspects beyond technology following the indications of the critical theory of technology developed by Feenberg [20]. Instead of illustrating the positive aspects of CPSs, which are fairly well known and advertised, we rather prefer to question the mainstream assumptions of technological neutrality that are narrowly evaluated by economic benefits. The main argument is that CPSs can be designed to not be subordinated to the commodity form, and thus, they can be used to govern the use of shared resources without market mediations by opening access and then creating new commons. Furthermore, the social impact of the new ICT technologies has been reviewed by discussing aspects related to data ownership, global platforms, fake news, and hybrid warfare. The list is not supposed to be extensive; it only serves as indications of the challenges of the society we live in, its limitations, and potentialities. We need to move beyond the capitalist realism [48] in order to build new ways of living with CPSs aiming at a future society based on sharing and without exploitation.

Exercises

11.1 Different governance models in CPS. Choose a CPS of your interest.
  (a) Demarcate the CPS based on its peculiar operation and its conditions of existence.
  (b) Indicate how the three types of governance models presented here can affect the selected CPS.
  (c) Identify one benefit and one drawback of the three governance models with respect to this CPS operation.
  (d) Write about how this CPS could be designed in a society that is not organized by commodities but by commons.

11.2 Thinking about the social impact of a CPS. Think about a smart watch that sets individual performance goals, measures heart beat, records how long the person stays active and sleeping, and also shows the most viewed news of the last hour. Discuss how this can be related to:
(a) data ownership;
(b) global platforms;
(c) fake news;
(d) hybrid war.

References

1 Cesarino L, Nardelli PHJ. The hidden hierarchy of far-right digital guerrilla warfare. Digital War. 2021 ;2:16–20. https://doi.org/10.1057/s42984-021-00032-3.

2 Marx K. Capital: A Critique of Political Economy. vol. 1. Progress Publishers. (Original work published 1867); 1887. Available at https://www.marxists.org/archive/marx/works/1867-c1/.

3 Rubin II. Essays on Marx's Theory of Value. vol. 23. Pattern Books; 2020.

4 Althusser L, Balibar E, Establet R, Ranciere J, Macherey P. Reading Capital: The Complete Edition. Verso; 2016.

5 Althusser L. On the Reproduction of Capitalism: Ideology and Ideological State Apparatuses. Verso; 2014.

6 Merriam-Webster Dictionary. Governance; 2021. Last accessed 29 September 2021. https://www.merriam-webster.com/dictionary/governance.

7 Facts NobelPrize org. Elinor Ostrom; 2021. Last accessed 29 September 2021. https://www.nobelprize.org/prizes/economic-sciences/2009/ostrom/facts/.

8 Ostrom E. Governing the Commons: The Evolution of Institutions for Collective Action. Cambridge University Press; 1990.

9 Ostrom E, Janssen MA, Anderies JM. Going beyond panaceas. Proceedings of the National Academy of Sciences of the United States of America. 2007;104(39):15176–15178.

10 Ostrom E. A general framework for analyzing sustainability of social-ecological systems. Science. 2009;325(5939):419–422.

11 Wolff RD, Resnick SA. Contending Economic Theories: Neoclassical, Keynesian, and Marxian. MIT Press; 2012.

12 Bak-Coleman JB, Alfano M, Barfuss W, Bergstrom CT, Centeno MA, Couzin ID, et al. Stewardship of global collective behavior. Proceedings of the National Academy of Sciences of the United States of America. 2021;118(27): e2025764118; https://doi.org/10.1073/pnas.2025764118.

13 Kühnlenz F, Nardelli PHJ, Karhinen S, Svento R. Implementing flexible demand: real-time price vs. market integration. Energy. 2018;149:550–565.

14 Dawson A. People's Power: Reclaiming the Energy Commons. OR Books New York; 2020.

15 Palensky P, Dietrich D. Demand side management: demand response, intelligent energy systems, and smart loads. IEEE Transactions on Industrial Informatics. 2011;7(3):381–388.

16 Arthur WB. Foundations of complexity economics. Nature Reviews Physics. 2021;3(2):136–145.

17 Blanchard BS, Fabrycky WJ. Systems Engineering and Analysis: Pearson New International Edition. Pearson Higher Ed; 2013.

18 Aglietta M. A Theory of Capitalist Regulation: The US Experience. vol. 28. Verso; 2000.

19 Bauwens M, Kostakis V, Pazaitis A. Peer to Peer: The Commons Manifesto. University of Westminster Press; 2019.

20 Feenberg A. Transforming Technology: A Critical Theory Revisited. Oxford University Press; 2002.

21 Nardelli PHJ, Alves H, Pinomaa A, Wahid S, Tomé MDC, Kosonen A, et al. Energy internet via packetized management: enabling technologies and deployment challenges. IEEE Access. 2019;7:16909–16924.

22 Giotitsas C, Nardelli PHJ, Kostakis V, Narayanan A. From private to public governance: the case for reconfiguring energy systems as a commons. Energy Research & Social Science. 2020;70:101737.

23 Nardelli PHJ, Hussain HM, Narayanan A, Yang Y. Virtual microgrid management via software-defined energy network for electricity sharing: benefits and challenges. IEEE Systems, Man, and Cybernetics Magazine. 2021;7(3):10–19.

24 Giotitsas C, Nardelli PHJ, Williamson S, Roos A, Pournaras E, Kostakis V. Energy governance as a commons: engineering alternative socio-technical configurations. Energy Research & Social Science. 2022;84: 102354. https://www.sciencedirect.com/science/article/pii/S221462962100445X.

25 Hussain HM, Narayanan A, Nardelli PHJ, Yang Y. What is energy internet? Concepts, technologies, and future directions. IEEE Access. 2020;8:183127–183145.

26 Chan F, Wong T, Chan L. Flexible job-shop scheduling problem under resource constraints. International Journal of Production Research. 2006;44(11):2071–2089.

27 Pournaras E. Collective learning: a 10-year Odyssey to human-centered distributed intelligence. In: 2020 IEEE International Conference on Autonomic Computing and Self-Organizing Systems (ACSOS). IEEE; 2020. p. 205–214.

28 Mashlakov A, Pournaras E, Nardelli PHJ, Honkapuro S. Decentralized cooperative scheduling of prosumer flexibility under forecast uncertainties. Applied Energy. 2021;290:116706.

29 Voigt P, Von dem Bussche A. The EU General Data Protection Regulation (GDPR). A Practical Guide, 1st Ed, Cham: Springer International Publishing. 2017;10:3152676.

30 Tikkinen-Piri C, Rohunen A, Markkula J. EU general data protection regulation: changes and implications for personal data collecting companies. Computer Law & Security Review. 2018;34(1):134–153.

31 Ullah M, Nardelli PHJ, Wolff A, Smolander K. Twenty-one key factors to choose an IoT platform: theoretical framework and its applications. IEEE Internet of Things Journal. 2020;7(10):10111–10119.

32 Ihlebæk KA, Sundet VS. Global platforms and asymmetrical power: industry dynamics and opportunities for policy change. New Media & Society. 2021;14614448211029662. https://doi.org/10.1177/14614448211029662

33 Wachter-Boettcher S. Technically Wrong: Sexist Apps, Biased Algorithms, and Other Threats of Toxic Tech. W.W. Norton & Company; 2017.

34 Karppi T, Nieborg DB. Facebook confessions: corporate abdication and Silicon Valley dystopianism. New Media & Society. 2020; 23(9):2634–2649. https://doi.org/10.1177/1461444820933549.

35 Lazer DM, Baum MA, Benkler Y, Berinsky AJ, Greenhill KM, Menczer F, et al. The science of fake news. Science. 2018;359(6380):1094–1096.

36 Seger E, Avin S, Pearson G, Briers M, Ó Heigeartaigh S, Bacon H. Tackling Threats to Informed Decision-Making in Democratic Societies: Promoting Epistemic Security in a Technologically-Advanced World. The Alan Turing Institute; 2020.

37 Umpleby SA. A history of the cybernetics movement in the United States. Journal of the Washington Academy of Sciences. 2005; 91(2): 54–66. https://www.jstor.org/stable/24531187.

38 Gerovitch S. From Newspeak to Cyberspeak: A History of Soviet Cybernetics. MIT Press; 2004.

39 Noble DF. Forces of Production: A Social History of Industrial Automation. Routledge; 2017.

40 Osinga FP. Science, Strategy and War: The Strategic Theory of John Boyd. Routledge; 2007.

41 Thomas T. Russia's reflexive control theory and the military. Journal of Slavic Military Studies. 2004;17(2):237–256.

42 Lefebvre V. Conflicting Structures. Leaf & Oaks Publishers; 2015.

43 Lepskiy V. Evolution of cybernetics: philosophical and methodological analysis. Kybernetes. 2018;47(2):249–261. https://doi.org/10.1108/K-03-2017-0120.

44 Vasara A. Theory of Reflexive Control: Origins, Evolution and Application in the Framework of Contemporary Russian Military Strategy. National Defence University; 2020.

45 Reichborn-Kjennerud E, Cullen P. What is hybrid warfare? NUPI Policy Brief; 2016.

46 Beer S, et al. Ten pints of Beer: the rationale of Stafford Beer's cybernetic books (1959–94). Kybernetes: The International Journal of Systems & Cybernetics 2000;29(5–6):558–572(15). https://doi.org/10.1108/03684920010333044.

47 Medina E. Cybernetic Revolutionaries: Technology and Politics in Allende's Chile. MIT Press; 2011.

48 Fisher M. Capitalist Realism: Is There No Alternative? John Hunt Publishing; 2009.

12

Closing Words

We started the book with the following questions:

> What is a cyber-physical system? Why should I study it? What are its relations to cybernetics, information theory, embedded systems, industrial automation, computer sciences, and even physics? Will cyber-physical systems be the seed of revolutions in industrial production and/or social relations? Is this book about theory or practice? Is it about mathematics, applied sciences, technology, or even philosophy?

If the task set by this text has been accomplished, the reader shall be capable of answering them all with full confidence. Now, let us move back and review what we have learned.

The rationale behind this book was to provide a still missing theory of cyber-physical systems (CPSs) beyond the particularities of specific study cases and their associated technological development. We followed the steps of cybernetics but carefully indicating the specifics of CPSs instead of using it as a universal framework. We began by reviewing the fundamentals of cybernetics, complexity sciences, and systems engineering (Chapters 1 and 2) to then explain the concepts of data and information that are fundamental to build the cyber domain of CPSs (Chapters 3 and 4). Graph theory was presented as the way to structurally characterize the relations between elements forming a network (Chapter 5) followed by an overview of different decision-making approaches (Chapter 6).

From those building blocks, a theory of CPSs can be proposed assuming a self-developing multiagent reflexive-active system constituted by three autonomous but interdependent layers, namely physical, data, and decision layers (Chapter 7). Each layer has its own "law" and limitations, but the dynamics of the CPS cannot be reduced to individual layer characterization alone; the theoretical understanding of CPSs requires characterization of the relations within one layer and along the three layers (Chapter 8).

Cyber-physical Systems: Theory, Methodology, and Applications, First Edition. Pedro H. J. Nardelli.
© 2022 The Institute of Electrical and Electronics Engineers, Inc. Published 2022 by John Wiley & Sons, Inc.

The proposed theory assumes specific enabling information and communication technologies (ICTs) (Chapter 9) in order to be used to design or study actual realizations of CPSs in the real world, including large-scale applications in energy systems and industrial plants (Chapter 10). This brings us to the social impact of CPSs in the existing capitalist society with global corporations dominating most of the ICT capabilities, leading to (geo)political and social reactions and associated movements; at the same time, CPSs open new opportunities to build a different mode of production based on commons (Chapter 11).

In what follows, a few items will be presented as suggested ways forward considering the world as in 2021. The first item is to emphasize that a strong theoretical foundation (as presented in this book) is needed to produce effective interventions. The second is a few selected theoretical and practical open challenges related to CPSs. The third focuses on what is nowadays called the fourth industrial revolution. The last one is a word of hope suggesting that CPSs could be designed to support peer production and manage shared resources to have a world free of exploitation.

12.1 Strong Theory Leads to Informed Practices

It is clear that so many CPSs are deployed already. However, as a researcher and teacher in this field, I was missing something. The (over)specialization of some theories and practices has resulted in a sort of myopic, too modular, understanding of reality. Few theoretical constructions like the ones from systems engineering or cybernetics offer tools to start looking at functional wholes. Other mathematical theories are transdisciplinary, offering quantifiable tools to assess and predict different phenomena. Examples of such theories are: probability theory, information theory, network sciences, game theory, optimization, dynamical systems, and control theory.

In the case of systems engineering, the main problem is that the different isolated parts are only schematically combined to perform joint tasks. The theory has a practical value and is not incorrect, but it provides too narrow a characterization of interrelated processes. Cybernetics, and specifically, cybernetic management based on viable system model (VSM) presented in the last chapter, offer options to systems engineering. However, it seems to me they have more a heuristic value than a scientific one because of their tendency of transposing concepts and theories from different scientific domains without the care they deserve. Cybernetics and some other organizational theories before it like Tektology of Alexander Bogdanov (1873–1928) are attempts to build universality of organizational structures regardless of the specificity of the objects. In fact, more than a science, cybernetic thinking is good at providing rules for actions toward a goal; in Wiener's words, cybernetics is related to *teleological mechanisms* in machines and animals. The case of mathematical theories is different: they are indeed scientific (consistent and always true) in their own domain with respect to their particular mathematical objects, axioms, and methods. Forcing mathematical truths to other domains usually results in inconsistencies and unsound results, despite the formalism.

In this book, my aim was to precisely define the object of the scientific discourse – the CPS and its three layers – by identifying and articulating its specific abstract forms considering the different mathematical theories that can provide scientific knowledge to design and assess different concrete CPS formations. In other words, the approach taken here was to present a strong foundational theory of the abstract object of CPS that can inform scientists, researchers, designers, and analysts in the field.

12.2 Open Challenges in CPSs

There are several open challenges in CPSs in both theory and practice. In theory, the foundations provided in this book allow further developments of how to properly characterize the relation between the structure of awareness and the structure of action of a given CPS, and how this coupling affects its self-development. Besides, we described how the proposed theory could be further developed considering the state of the art in networked control theory that explicitly includes communication and computation aspects in the dynamics of physical systems. Likewise, the studies of multiagent systems with interactive decision-making processes should be revisited considering the three-layer model. The information and communication theories for CPS might also be extended to include semantic and functional aspects related to the CPS itself, as part of its conditions of reproduction or its operating conditions, where data and communications are not agnostic: data are acquired from somewhere that gives meaning (semantics) and are used to inform some process that has a function (functional). These are only a few examples, and the list could be extended indefinitely.

In practical terms, there are also very different challenges, from providing ultra low latency and reliability end-to-end communications for machines to distributed methods to coordinate swarms of heterogeneous robots performing joint tasks. Another important research direction refers to the development of explainable predictive machine learning methods, making their outcomes meaningful for end users. The management of data networks, energy consumption of methods involving higher level data processes like distributed ledgers, integration of distributed energy sources to supply CPSs, radio resource allocation for wireless communications, and software development for systems with heterogeneous decision-makers and agents are but a few examples of practical technological developments that are necessary to deploy CPSs. Besides, there are also several other aspects beyond technology already discussed in the previous chapter.

12.3 CPSs and the Fourth Industrial Revolution

The capitalism is frequently divided into periods related to industrial revolutions, which are large-scale changes in the organization of productive forces and in their effects in the relations of production. The first industrial revolution starting at the end of the eighteenth century is associated with the mechanization of industrial activities. The second industrial revolution starting at the end of the nineteenth century refers to technological developments of internal combustion engines, chemical synthesis, electricity grids, and new methods for telecommunication, as well as new products like automobiles and airplanes in the early twentieth century; these are associated with mass production. The third industrial revolution starting during the second half of the twentieth century refers to the automation of specific tasks enabled by the development of specialized ICTs and also include the development of nuclear energy and the further industrialization of activities and practices. The fourth industrial revolution is associated with the Internet of Things, interactive robots, digital twins, and the like: in summary, CPSs. Figure 12.1 depicts this usual classification.

Although this representation might not be accurate enough, it describes the historical tendency. Moreover, Figure 12.1 can be seen as a kind of self-fulfilling prophecy indicating the path that the industrial research and development ought to follow in order to reach an advantageous position in the global market. In any case, all four types of industries associated with the four revolutions currently coexist, the ones from the second and third ones seeming to be the dominant today. What is remarkable is that the move from the second to the third and fourth revolutions implies a decrease in direct working force at the factory floor, which leads

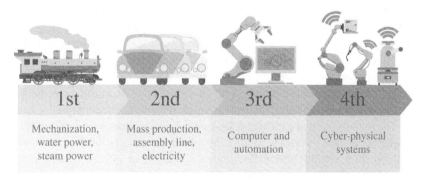

Figure 12.1 From the first to the fourth industrial revolution.

to unemployment of the traditional working classes. At the same time, the rate of exploitation tends to grow, shifting to different sectors of economy like software production. The consequences are concrete, but very little can be said with certainty today as far as it concerns current economic, (geo)political, and social struggles, the results of which are open and unknown.

12.4 Building the Future

My last word is a word of hope. Science fiction literature has narrated different utopian and dystopian futures, and in both cases, ICTs are pervasive to create cyber mediations anywhere, anytime (as the members of wireless communication community like to pitch and preach). Our daily life seems pretty much a (dystopian) future based on full commodification of every physical and cyber aspect of natural, individual, and social existence combined with centralized algorithmic governance approaches based on badges, nudges, and rewards to guide individual behavior. This phenomenon might be called gamification of life. In this world, the commander is an abstract entity with concrete effects, whose only goal is its perpetual self-growth toward unlimited accumulation; its name is Capital. Against this, the only hope is to work to construct a future where the existence is transindividual and shared; this is only possible when the commodity form disappears and a new mode of production based on commons is established. As indicated in the previous chapter, CPSs may play a major role in transforming our current society to establish a new one based on sharing. If this is what has to be accomplished, we need to be ready to produce theoretical and practical knowledge that will construct CPSs that are designed exclusively for the commons.

Exercises

12.1 **CPSs: theory, methodology, and applications.** Write one paragraph summarizing each 1 of the 12 chapters of this book.

12.2 **Thinking the future.** Write a two-page piece of science fiction describing a vision of the world where CPSs are pervasive.

Index

Cyber-physical Systems: Theory, Methodology, and Applications, First Edition. Pedro H. J. Nardelli.
© 2022 The Institute of Electrical and Electronics Engineers, Inc. Published 2022 by John Wiley & Sons, Inc.